BuiLT BY OiL

BUILT BY OIL

Gwilym Roberts

David Fowler

Ithaca
PRESS

Built by Oil

Ithaca Press is an imprint of Garnet Publishing Limited
Published by Garnet Publishing Limited
8 Southern Court
South Street
Reading
Berkshire RG1 4QS UK

First edition

ISBN 0 86372 189 3

British Library Cataloguing-in-Publication Data.
A catalogue record for this book is available from
the British Library.

Jacket design by David Rose
Book design by David Rose
Typeset by Samantha Abley

Printed in Lebanon

CONTENTS

CONTENTS

FOREWORD

Since my first visit to Kuwait in 1952 until my retirement 40 years later, I have worked continuously in the Middle East. Throughout this period I have become increasingly aware of the magnitude and impact of the projects that have been designed and constructed in the region since the discovery and exploitation of its oil reserves.

In virtually every Middle Eastern country engineers have created whole new infrastructures which, by enabling the benefits of modern technology to be incorporated into the context of ancient civilisations, have advanced and, in most cases, transformed the living standards of the citizens.

The effect has been comparable to that of Victorian engineers in Britain a century and more ago when, through the construction of the country's transportation, water, and sewerage infrastructure, they fostered the industrial revolution and enabled our cities to be created, thereby establishing the basis for our modern civilisation.

On my first visit to Kuwait, water was imported by dhow from the Shatt al-Arab river and distributed in goatskins on the backs of donkeys, and I was involved in designing the state's first piped fresh water scheme, which had a capacity of a million gallons a day. Nearly 40 years later the Americans bombed a water pumping station which my company had recently commissioned and which had increased the country's water capacity to about 200 million gallons a day (a statistic which is all the more remarkable when it is remembered that the country's rainfall is minimal and that all fresh water has to be distilled from the sea). Such has been the tempo of change.

Because of the UK's historical and political connections with many of the countries in the region (often the result of the need to safeguard the trade route to India), the various rulers turned first to British engineers for advice when oil revenues started to flow and they were able to consider the development and modernisation of their countries.

In consequence, British engineering consultants have played a major role in developing and designing the various projects that have been constructed in the Middle East over the past 40 years. More recently, however, engineers of other nationalities have participated to an increasing extent, sometimes in multinational consortia such as on the massive Greater Cairo Wastewater Project where American, British and Egyptian engineers have now been joined by Italians. British dominance has not been as pronounced in construction and equipment supply, which has been undertaken by a wide range of contractors from across Europe and Asia.

This book, in whose compilation I have been very greatly helped by my co-author David Fowler, is an attempt to record and describe some of the more significant projects that have been undertaken in the Middle East during the past half century, and to set these within the historical and political contexts of the various countries.

Naturally, we have had to be selective about which projects to describe, and while we have tried to be objective in choosing the most significant, in the end a degree of personal judgement entered in. Space has also prevented us, in general, from mentioning by name the local contractors and consultants who worked in partnership with the expatriate firms and made an invaluable contribution to the work. That contribution is nonetheless recognised.

That, indeed, is the most important purpose of this book: it is a personal tribute to the tens of thousands of engineers and others from all walks of life and of many nationalities, whose dedication and hard work have enabled achievements of the scale and magnitude described to have been conceived, designed and built.

Gwilym Roberts
Westmeston, Sussex
January 1995

ABBREVIATIONS

b/d	barrels a day
dwt	deadweight tonne
ft	feet
g/d	gallons a day
ha	hectare (s)
hp	horsepower
km	kilometre (s)
km/h	kilometres per hour
kN/m^2	kilonewton/m^2
kVA	kilovolt-amp
l/d	litres per day
l/h	litres per hour
l/s	litres per second
m	metre (s)
mg/l	milligrams per litre
m/s	metres per second
m^2	square metre (s)
m^3	cubic metres
m^3/s	cubic metres per second
m^3/d	cubic metres per day
m^3/h	cubic metres per hour
m^3/y	cubic metres per year
mm	millimetre (s)
mph	miles per hour
MW	megawatt
N/mm^2	newton/mm^2
ppm	parts per million
t	tonne(s)
t/d	tonnes per day
t/h	tonnes per hour
t/year	tonnes per year
TWh/y	Terawatt hours per year

MERGERS

Over the years a number of firms have merged. They are referred to in the text by the name by which they were known at the time being written about. To save repetition we list here those companies which occur frequently in the text and the name changes they went through.

Acer formed from Freeman Fox & Partners John Taylor & Sons in 1987.

Sir Bruce White, Wolfe Barry & Partners joined the group in 1991.

Coode & Partners merged with Lemon & Blizard in 1986 to form Coode Blizard. This practice merged with Graham Consultants late in 1994.

Montgomery Watson formed from Watson Hawksley and James M. Montgomery (1993)

Watson Hawksley was formed when J. D. & D. M. Watson and T. & C. Hawksley merged in 1978.

Mott MacDonald Group formed from Mott, Hay & Anderson and Sir M. MacDonald & Partners in 1989. Parsons, Brown & Newton had become part of Mott, Hay and Anderson in 1983.

Ewbank Preece, formed from Ewbank & Partners and Preece Cardew & Rider in 1982, joined the group in 1994.

Pencol combined with sister company Spencer & Partners to form Penspen in 1990.

MAP SHOWING
THE COUNTRIES OF
THE MIDDLE EAST

—

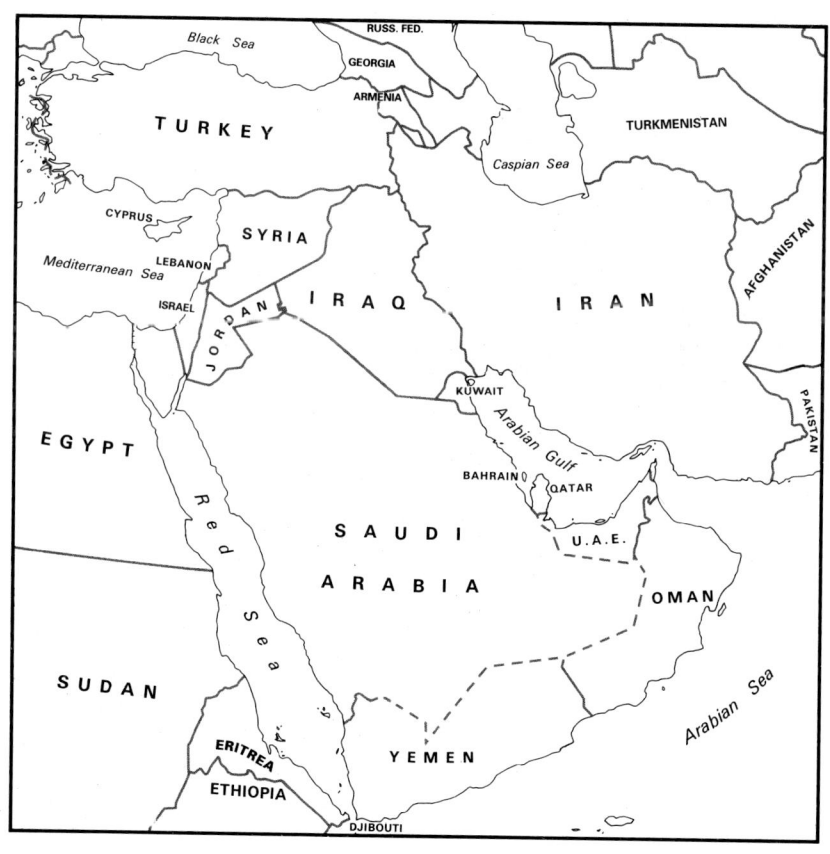

ACKNOWLEDGEMENTS

―――

The authors would like to thank the following people who helped with their research:

Peter Banks, David Trumper, Hamish Somerville, Derek Wolstenholme, Amanda Powell-Smith and Pat Murtagh (Acer); John Ward, Ron Marsh (Ove Arup); Col Charles Peacey (Association of Consulting Engineers); David Kell, Annie Robinson (Binnie); BK Bardhan-Roy (Jan Bobrowski & Partners); Margaret Brown (Coode Blizzard); Tim Westman (Costain); John Allen, John Bowcock, Brian Brent, Geoff Brice, David Greatbatch, Roger Harman, Peter Hunter, Bill Pirie, Robin Snaddon, Tim Woods Ballard (Gibb); Paul Wilkinson, Susan Pacey (Halcrow); Mike Chrimes, Robert Thomas, and the Institution of Civil Engineers (ICE) Library staff; Bruce Boys, Bruce Foot, Julie Mitchell, Graham Read (Laing); Shakir Al Kubaisi, Peter Sockett, Jane Sykes (Maunsell); Peter Lawrence, Mike Thompson, Alan Willson, Isobel Baird, Jon Balley (Montgomery Watson); Jill Burger (Mouchel); Mike Blackburn, Mike Blyth, Mike Burley, David Donald, David Elliot, Mike George, Brian Jackson, Bernard Morphew, John Pavey, Jim Perry, Rex Vickers, Dick Volmer, Andy Webb, Teresa Henry, Anne Buttfield (Mott MacDonald); the past and present staff of *New Civil Engineer*; Mrs M. Magnusson (Royal Engineers' Library); Gordon Matthews, Ron Sweetapple, Kempton Bedell-Harper (Rendel Palmer & Tritton); Professor Tony Allan (School of Oriental & African Studies, University of London); Colin Holmes, Richard Cooper (Scott Wilson Kirkpatrick); Tony Wheel, John Hawkins (Tarmac); Colin Clark, Chris Irwin-Childs, Mike Kelly, John Cox, Mickey Lochhab (Taylor Woodrow); Anthony Hasluck (Trafalgar House); Peter Grey, Bruce Snelling (Wimpey); Roger Dobson, Peter Hipwell, Viv Hoad, Cecil Rogers; and the late Geoffrey Coates, formerly Chairman of Sir Alexander Gibb & Partners and of FIDIC, the International Federation of Consulting Engineers, who epitomised the postwar generation of engineers working in the Middle East and who sadly died shortly after contributing his recollections.

A database listing salient details of a comprehensive range of Middle East projects, collected by the authors while researching this book, has been lodged with the ICE Library, Great George Street, London, SW1.

INTRODUCTION
AND HISTORICAL
BACKGROUND

When, some 10,000 years ago, nomadic tribesmen first developed pastoral and arable farming techniques along the banks of the River Huang He in China, the Indus in Pakistan, the Tigris and Euphrates in Iraq and the Nile in Egypt, the seeds of our modern civilisation were sown.

Long-term survival in such arid climates was possible only because the contemporary engineers developed irrigation networks. From that time onwards the artefacts and structures created by engineers have advanced mankind's living standards having provided the foundation for the well-being and prosperity of each successive civilisation.

Of the four cradles of civilisation, those in the Middle East are best known to us. This is partly because of proximity; partly because it is from these that the Persian, Greek, Roman and Arab Empires grew and from which in turn our own modern western civilisation developed; and partly because the Egyptians had greater access to stone which means that more of the structures built by them are still standing today.

Each of the early civilisations, however, created engineering works of a size, variety and complexity which are impressive even by today's standards. Their achievements are particularly amazing in view of the knowledge and equipment available to their builders.

Sir Harold Hartley, in his keynote address on the subject of 'Engineering and Civilisation' given on the occasion of the 150th anniversary of the UK's Institution of Civil Engineers (ICE), noted that six of the Seven Wonders of the World were the work of engineers.[1] They included the Great Pyramid of Giza (which remained the tallest structure in the world until 1307 AD when it was surpassed by the central spire of Lincoln Cathedral, which however collapsed in 1548); the Lighthouse of Alexandria; and the Hanging Gardens of Babylon, which included a complex device for raising vast quantities of waters from the Euphrates to a considerable height. Only the statue of Zeus at Olympia was a pure work of art rather than an engineering feat.

At the 1955 Graham Clarke Lecture a paper entitled 'The Engineer's Contribution to the Conservation of Natural Resources', Sir Harold said:

> Sennacherib built a great aqueduct 50 miles long to bring water to Nineveh and to irrigate the orchards and gardens outside the city wall. It was as wide as an arterial road and paved with masonry; a dam with sluice gates regulated the flow and enabled water to be stored. These early irrigation engineers were most skilful in the design and control of their large canal systems. Some, like the Nahrwan canal running for 200 miles parallel with the Tigris, rivalled the largest modern works in their boldness of conception. They knew how to divert rivers by dams and to dig great artificial canals to lead the rivers into their irrigation systems and to avoid the risks of erosion and silt.[2]

Water was, of course, the key to life and prosperity. In addition to irrigation works, dams were built as early as 3000 BC in Egypt and Yemen; Nileometers which recorded the flood level of the Nile were used to determine annual taxes, while in the lands that are now Iran and Saudi Arabia many *qanats* (tunnels laid at a gradient to convey water to towns) were constructed over considerable distances.

Communication by land required the construction of highways and bridges, of which the most spectacular was the stone bridge across the Euphrates built for the Chaldeans at Babylon.

Maritime and river transport required the construction of harbours and ships, as well as bypasses for the Nile cataracts and a canal linking the Red Sea to the Mediterranean which was a precursor of the modern Suez Canal.

Towns were laid out and drains provided to cope with the flash floods that occasionally occurred. Archaeologists excavating at Mari in eastern Syria recorded:

> As for the drainage, it was effected by means of brick gutters laid under the pavement and of bitumen-lined clay pipes going down 30 feet underground. The whole system had been so skilfully planned and installed that the waters of a violent rainstorm, which burst one day during the excavations, were evacuated within a few hours, the drains having worked again, most efficiently, after forty centuries of disuse.[3]

Ctesiphon to the south of Baghdad with its huge vaulted arch, the pyramids and temples in Egypt, and the many other structures built throughout the Middle East millennia ago, are impressive not only for their sheer size, but also for the technical skill and human management that must have been needed to construct them.

Modern engineers who have worked in the Middle East in recent decades have thus not only had magnificent examples to follow, but have also had the privilege of continuing a magnificent tradition.

What exactly is meant by 'the Middle East'? The term, although European in origin, has gained acceptance even in the region itself, and has superseded the expression 'Near East' which was formerly used to describe Turkey, the Balkans, the Levant and Egypt. The definition used by the authors of this book comprises the Arabic-speaking world from Libya eastwards, plus Iran. Excluded are Pakistan, Afghanistan and the Mahgreb states of North Africa – Tunisia, Algeria and Morocco. Turkey has been included, even though it is strictly outside this definition and increasingly considers itself part of Europe, because of its historical links with the region through the Ottoman empire, and its geographical position as the gateway to the Middle East.

The Middle East, and particularly the fertile crescent which stretches from Basra in Iraq to Lebanon and along the eastern Mediterranean to the Nile valley, was settled, invaded and re-invaded throughout history as successive empires rose and fell. Arab, Greek, Roman, Byzantine and Ottoman rule came and went as the centuries progressed. But two of the empires which held sway in the region hold particular significance. The first was the Arab/Islamic empire founded in the seventh century AD; the second was the Turkish Ottoman empire established seven hundred years later.

It was early in the seventh century that the Prophet Muhammad, the founder of Islam, began his mission. At that time the Arabian peninsula was populated mostly by tribal peoples who had no written law and who worshipped a variety of spirits. Muhammad preached that there was one true God, who has revealed His will through prophets, Muhammad being the last of the line of prophets. He was also a skilled political leader. The Koran, which Muslims believe to be the literal word of God, together with Muhammad's teachings and actions, form a complete system of ideal morality in all spheres of life; in contrast with Christianity, there is no concept of a secular state distinct from the faith. Within thirty years the new religion had spread over most of the Arabian peninsula, and under its influence the scattered tribes coalesced.

Muhammad's immediate successors built on this foundation to challenge the two great empires of the day, those of Byzantium and Persia, and to create an Arab empire which covered North Africa, southern Spain, Portugal, Iraq, Syria and Egypt. Islam spread widely throughout this area, and also in Asia Minor (Turkey) which the Arabs did not conquer.

Damascus was the early capital of the empire; after a century, however, the centre of power shifted to Baghdad which became a vast free trading centre. The Arabs assimilated the practices of the powers they overwhelmed, such as their systems of government, and developed them. Learning also flourished in the two centuries before the break-up of the Arab empire which could not, in the long run, be held together from the centre.

The Ottoman, or Osmanli, empire was founded by Osman, a Turkish warrior prince, and his descendants. They first conquered the Balkans and eventually overwhelmed the remnants of the Byzantine empire, including its Middle Eastern possessions. For a considerable time the Ottomans threatened Europe before turning their attention instead to Asia. Of the Arabic-speaking world, only Morocco, Oman, and the central area of the Arabian peninsula remained unconquered. Like their predecessors in the first Islamic empire the Ottomans showed tolerance towards the people of the lands they conquered, and to their religious beliefs. But they proved less adept at learning from, and absorbing, outside influences.

Though the Ottoman empire lasted five centuries, its decline began less than halfway through this period. By the nineteenth century, competition between the emerging powers of Britain, France and Russia conspired to help it survive longer than it might otherwise have done: each preferred a weak Ottoman empire to a strong breakaway group, influenced by a rival power, and having the potential to disrupt trade.

Among the causes of the Ottoman empire's decline was its disdain for industry and commerce which allowed it to be overtaken by the growing industrial economies of west Europe. The Middle East, whose economy in the Middle Ages had been based on commerce, became instead dependent on subsistence agriculture. At the same time its political system, which was feudal but non-hereditary, made for an egalitarian society but prevented the emergence of a powerful nobility on the European model. Meanwhile, as the west European economies grew prosperous, so their military strength came to surpass that of the Ottomans.

Attempts to reform the empire proved fruitless. Muhammad Ali, who became ruler of Egypt in the early nineteenth century, tried to encourage industry there as part of his plans for a more modern Arab empire. He conquered Sudan and Syria and eventually became strong enough to challenge the Sultan of the Ottoman empire in Istanbul. At this point western powers led by Britain decided that Muhammad Ali's influence was becoming too great, and helped the Sultan to dispossess him of all his conquests except Egypt. Muhammad Ali's industrialisation plan failed partly because it had been overwhelmingly biased towards aiding his military ambitions, and partly because the country lacked readily available sources of power.

In Istanbul, in the mid-nineteenth century, a reform programme overseen by Mustafa Reshid Pasha introduced a code of secular law and began a programme of improvements to the education system, but these were undermined by the empire's relative economic weakness.

It was not until the Middle East acquired its vast oil wealth that it was able to gain the trappings of industrialised west European civilisation. The emergence of a distinct class of entrepreneurs also had to wait until that time.

The growth of civilisations depends on certain common factors. For civilisation to develop, five groups of people are necessary: scientists, engineers, statesmen, merchants and educationalists working in collaboration.[4] But it can be argued that it is the engineers who lay the essential foundations for the community's prosperity.

Of overriding importance is a supply of water, essential for life and for agriculture. It is thus no coincidence that all the early civilisations grew up near rivers. Water supply, then, is the first concern of engineers in civilised society. Water is a catalyst which allows communities to become established; a safe water supply also ensures the health of the community. Once this is provided, the engineer's attention turns to the infrastructure which will allow trade to prosper and the economy to develop: creation of cities, establishment and improvement of communications and development of sources of power. The network of roads across the Roman empire is an obvious example of how engineers work to establish and develop a civilisation.

The development of such an infrastructure allows merchants and traders to create wealth through commerce, while politicians and administrators play their part by developing a strong state and taking responsibility for its security.

Engineers' concerns were essentially the same in Victorian Britain. Though it was the highly visible aspects of engineering, such as the railways, which captured the public imagination, perhaps the most important development of the age was the discovery that diseases like cholera were water-borne. Great water engineers of the time included James Simpson, who moved the water intakes of the Lambeth water company up river to Long Ditton and away from the polluted waters of the Thames in central London, and Sir Joseph Bazalgette, who rebuilt London's sewerage system. Between 1850 and 1870, cholera virtually disappeared as a cause of death in Britain. Real economic progress would not have been possible had this scourge not been conquered.

Simultaneously, 'railway mania' was transforming communications, drastically reducing journey times to even the far-flung corners of the island, and eclipsing canals as a means of transporting freight. Manufacturing industry also developed apace, supplied by new sources of energy – steam, founded on a vast wealth of coal, and hydraulic power. These sources of power began to allow machinery to take the place of human labour in all activities. The 'steam navvy' revolutionised the excavation of the Manchester Ship Canal, for example.

This pattern can also be discerned at work in the Middle East in the second half of the twentieth century. Concessions to prospect for, and develop, the area's oil reserves were granted to western organisations and initial efforts were necessarily directed towards providing the means to extract and refine the oil and transport it to its markets. But once the oil wealth began to flow, the comprehensive development of infrastructure became the primary aim of the oil-rich nations.

New and expanded ports allowed trade with the west to increase; new roads linked centres of population; power stations were built to provide energy for new factories; airports were constructed to cope with the huge increase in business travellers. In Jubail in Saudi Arabia, an industrial complex on the scale of Teesside in north-east England was built from scratch in 10 years. Many Arab states embarked on an ambitious series of five-year plans covering all aspects of the country's infrastructure from ports and roads to universities and hospitals.

At the same time, city populations grew at an unprecedented rate as local people migrated there and expatriates were brought in to work in the new industries and on construction projects. Water supply and sanitation were once again top priorities.

Another crucial factor in the development of the Middle East was its strategic position. To control the eastern Mediterranean and the Gulf was to control trade routes from Europe to Asia and the Far East. Following the reign of Augustus in the first century AD the Romans cleared the Red Sea of pirates in order to allow trade to India. In 1839, worried by Muhammad Ali's growing influence along the east coast of the Red Sea, Britain seized the port of Aden on the pretext that the Sultan of Lahej, in whose territory the port lay, had permitted British shipping to be molested. Further east, to protect its trading ships against piracy, Britain entered into maritime truces in 1820 and 1853 with the sheikhdoms along the Gulf, under which these 'Trucial States' – now the United Arab Emirates – became British protectorates.

In the eighteenth and nineteenth centuries the overland route from Beirut and Damascus via Baghdad and Basra was the fastest way of transporting goods and mail to and from India and was of vital importance to Britain's economy. For this reason Britain's interests in Iraq were the responsibility of the Indian Office rather than the colonial office, and Indian rupees were the currency in Kuwait.

Britain was not the only nation seeking to expand its influence in the region. During the reign of Muhammad Ali in Egypt close links had been established with the French. Muhammad Ali's successor, Abbas, was surprisingly sympathetic to Britain, and allowed Robert Stephenson, son of the pioneering British railway engineer George Stephenson, to build the Cairo–Alexandria railway in 1850–51. Said, who succeeded Abbas in 1854, again favoured France. He embarked on a vast programme of public works which so improved the country's infrastructure that he boasted that it was now part of Europe rather than Africa. Unfortunately this progress entailed the amassing of a mountain of debt which eventually led to Egypt's bankruptcy and Said's downfall. This was not, however, before Said had granted a concession to the French engineer Ferdinand de Lesseps to build the Suez Canal.

The idea of a canal, especially one under French control, was anathema to Britain who would have preferred the Cairo–Alexandria railway to be extended to Suez. Thus, when in 1875 Egypt's crippling debt burden forced Said to sell his 44 per cent shareholding in the Suez Canal Company, the British prime minister Disraeli was quick to acquire it – to the chagrin of the French.

Eventually Egypt's financial state became so parlous that Anglo-French control was imposed on the country to raise taxes in order to pay off the debt. So, to protect its trading interests, Britain developed close links with the Middle East which, fortuitously, laid the foundations for British participation, decades later, in the development of oil in the Gulf, and the subsequent construction boom there.

Other important transport links subsequently developed in the region and influenced its later growth. The route of the Orient Express was extended from Scutari, opposite Istanbul on the Bosphorus, to Baghdad and Basra, by German engineers. When the age of air travel began, Imperial Airways (later British Overseas Airways Corporation and then British Airways) used Basra, Bahrain and Dubai as stopover points for its flying-boat services. The Nairn Bus Company, meanwhile set up after World War I by two New Zealander brothers, ran services from Beirut and Damascus to Baghdad.

But the strategic importance of the Middle East increased enormously with the discovery of oil. In 1901, the Shah of Persia had granted the Englishman William Knox D'Arcy a 60-year concession to search for oil throughout the Persian empire. D'Arcy was unsuccessful for several years, and had almost run out of funds when the First Sea Lord, Admiral John Fisher, persuaded the Burmah Oil Company to supply further investment. Fisher believed that converting the British fleet to burn oil would increase its fighting power by 50 per cent. But most of the world's oil was produced by the US or Russia, and Britain needed an independent supply. In 1908, when D'Arcy was about to give up, he struck oil. The Anglo-Persian Oil Company (later called British Petroleum) was formed to exploit the discovery.

Persia remained the only oil-producing country in the Middle East until the end of the 1920s. An oilfield was subsequently discovered near Kirkuk in Iraq in 1927, which started to bring in substantial revenue by the mid-1930s. When a commercial quantity of oil was found in Bahrain in 1932, after only six months' drilling, the vast potential oil wealth of the region became apparent.

Concessions to explore followed in Saudi Arabia, Kuwait and Qatar which resulted in discoveries shortly before World War II, but development of these fields could not begin until after the war.

Concession agreements had not always been cast in the most favourable terms to the countries granting them, and oil revenues were in some cases slow to build up. Nevertheless, the stage was now set for the transformation of the Middle East in the second half of the twentieth century. This transformation would be on a par with that which took place in Victorian England, allowing the region in which civilisation first began to regain the ground it had lost to European economies in the previous five centuries. Engineers, and in particular British civil engineers, were at the forefront of that transformation.

IRAN

Iran was the first Gulf state in which oil was discovered and exploited. This had disadvantages as well as advantages for it meant that it was also the first state to have to battle with oil companies over the meagre share of profits it was initially receiving, as a result of which an international boycott of Iranian oil was temporarily imposed during the early 1950s.

But having benefited from oil since early this century, Iran's economy was more developed than those of most of its Gulf neighbours by the early 1970s. When the quadrupling of oil prices following the Arab–Israeli war of 1973 brought an enormous windfall to all the oil-producing states, Iran's response was to embark on a massive programme of infrastructure development and industrialisation. A similar strategy was also followed with considerable success by Saudi Arabia. In theory Iran should have been in a good position to develop a strong industrial economy and become a major force in the region. Instead, ironically, the dash for industrialisation helped to sow the seeds of the 1979 revolution and the shah's replacement with Ayatollah Khomeini.

Iran, or Persia as it was formerly known, has been a Muslim country since the seventh century AD. Most of the population were Sunni Muslims until the sixteenth century. During the Safavid empire, to help engender a sense of national identity during the struggle with the Ottomans, Shiism was made the official rite and has remained so ever since.

During the nineteenth century Britain and Russia competed for influence in Persia. The ruling Qajar dynasty was extravagant but, as the country's economy was undeveloped, sources of income were restricted. One solution to the income dilemma was to grant concessions to exploit mineral wealth. A 60-year concession to prospect for oil was thus granted to the Englishman, William Knox D'Arcy, in 1901.

The search for oil was not easy and came near to the point of being abandoned more than once before oil was finally struck at Masjid-i-Sulaiman in 1908. The Anglo-Persian Oil Company (APOC), later known as the Anglo-Iranian Oil Company (AIOC) and which finally became British Petroleum, was formed to exploit the discovery.

The country's economy remained chaotic, and after World War I British advisers were sent in to assist the government. In 1921 the Qajar dynasty was overthrown in a coup by Reza Khan, or Reza Shah Pahlavi to give him his imperial title, an officer in the Cossack Brigade. He carried through some reforms which included curbing the influence of the clergy by making the legal and education systems secular. He also created a disciplined army. Initially his strong leadership and nationalism were welcomed, but his later attempts to modernise society on western lines, including the banning of traditional clothes, caused resentment and confrontation.

During World War II, when Iran was occupied and used as a western supply route to the Soviet Union, Reza Shah was forced to abdicate in favour of his son, Muhammad Reza, who was to reign until 1979.

After the war, though British troops withdrew from Iran, British influence remained strong because of the role of the AIOC. Britain and the AIOC became targets of Iranian nationalism, for which the veteran politician Muhammad Mossadegh became principal spokesman. An ambitious seven-year development plan was launched in 1949. But when it became clear that, while the oil company's revenues had increased tenfold between 1940 and 1955, Iran's share had increased only fourfold, a clamour for the nationalisation of oil arose.

The shah was obliged to appoint Mossadegh as prime minister and to give assent to a nationalisation bill. The AIOC withdrew from the country and an international boycott of Iranian oil was organised, slowing the country's revenues to a trickle. The shah attempted to replace Mossadegh as prime minister but his challenge failed and he fled the country. Within days, however, Mossadegh was deposed in a coup organised by the US Central Intelligence Agency (CIA) and the shah returned.

From that time the US became the main western influence on Iran. A new oil agreement was hammered out with a consortium of US, British, French and Dutch companies. These events opened the way for the shah to create a dictatorship which was to last for 25 years.

This period was initially characterised by extravagant military spending and economic mismanagement. But the only serious challenge to the shah came in the early 1960s from Ayatollah Khomeini, who was imprisoned and then exiled. The late 1960s and early 1970s were a period of stability.

The fourfold increase in oil prices in 1973 seemed to provide the shah with the opportunity to realise his greatest ambition, that of making Iran a major power in the Middle East. He boasted, somewhat unrealistically, that the country would be one of the six most advanced industrial countries in the world by the end of the century, a Japan of western Asia. In fact many people in the country were still living in conditions akin to those in the poor developing countries.

Estimates for the cost of Iran's fifth five-year plan, to run from 1973 to 1978, were doubled to £30,000 million, and the whole of the original version of the plan was compressed into the first year of the new one. But implementation of the plan began to impose intolerable strains on the country's infrastructure and social fabric. There were two main differences between the experiences of Iran and Saudi Arabia in this respect. First, in Iran too little emphasis was attached to developing the infrastructure in advance of industrialisation. By 1975, roads were overcrowded, hotels were booked a year in advance and the telephone system was strained beyond capacity. By 1976, there was a 225-day delay at the main port of Khorramshahr. The economy overheated and inflation reached 40 per cent.

But there was another, more important, difference. Whereas Saudi Arabia had adopted a policy of modernising the country while keeping its traditions intact, Iran attempted to abandon its past. Women were encouraged to abandon traditional dress and go out to work. Agriculture was to be modernised to free millions to work in urban centres – in fact an exodus of farm workers to towns merely dislocated the farming system. And while the new middle classes were alienated by the rise in inflation, the vast majority of the population reacted against the wholesale, hasty westernisation, which they saw as un-Islamic and evil.
By 1976 oil revenue had dropped and plans were cut back – but this would have had to happen in any case because infrastructure would have proved a limiting constraint if cash had not.

The shah had generally been able to suppress dissent throughout his reign. But this time he underestimated its extent and from then on the opposition began inexorably to gain momentum. From exile in Paris Ayatollah Khomeini called for the shah's abdication. As the only figure opposed to the shah who, because he had been in exile, had never been tainted by compromise with the shah, he was idealised as the dissenters' leader. The shah vacillated between half-hearted concessions to reform and attempts to quell discontent by force. But eventually it became clear that he lacked the will for the extreme measures which would have been needed to restore his power. In January 1979 he flew into exile. Two weeks later Ayatollah Khomeini returned to declare an Islamic republic.

Khomeini's authority was not unquestioned. He faced opposition from nationalists, left-wingers and senior figures in the clergy, but he had sufficient popular support to overcome this. Nevertheless the early years of the new regime were a period of uncertainty which at times came close to civil war. Neighbouring Arab states were worried about the implications of Iran's revolution. In 1980 Iraq took advantage of Arab anxieties to invade Iran.

Iraq invaded the oil-producing region of Khuzestan and made substantial territorial gains. However, the Iraqi leader Saddam Hussein had underestimated the morale of the Iranian forces and within a year the latter had recaptured most of the territory. There followed a five-year war of attrition, with considerable loss of life, rocket attacks and shelling of major cities on both sides, but with neither side making any significant gains. In 1986 Iran captured the Fao peninsula but failed to take Basra; a year later the United Nations called for a cease-fire, which was ignored. The Iraqis recaptured Fao and in 1988 Iran was forced to accept the cease-fire. The inconclusive war seriously disrupted oil production and caused considerable harm to both countries' economies.

Accepting the cease-fire meant considerable loss of face for the Khomeini regime but did not seriously threaten it. This, and the smooth transfer of power which took place on Khomeini's death, have demonstrated that the structures of the new republic are secure.

The fortunes of construction firms in Iran have been mixed. Iran has an indigenous professional class which was, however, overwhelmed by the scale of the 1970s' economic boom. This presented overseas firms with a wealth of opportunities. British consultants such as Ove Arup & Partners and Sir Alexander Gibb & Partners enjoyed considerable success there – from 1946 onwards in Gibb's case. Contractors, however, always found the market more difficult to break into.

Western firms left the country *en masse* at the time of the Islamic revolution and few have returned. The UK consultants John Taylor & Sons maintained links with regard to the Tehran sewerage project during the Iran–Iraq War and the project is now active again. Work also continued almost uninterrupted on construction of the Lar dam. But contractor Marples Ridgway, which was midway through a 300-km road contract, eventually abandoned work after the Iranian authorities stopped paying it.

A more open policy towards overseas firms adopted in 1991 raised hopes, but again economic mismanagement and lower than expected oil revenue have left Iran in debt. Many projects – including a programme of dam building in which Mott MacDonald is involved – have been cut back.

Railways

Railway construction was one of the few areas of civil engineering in which the Middle East boom did not cause a huge upsurge of work. An exception, the planned modernisation of the Tehran to Tabriz railway, fell foul of the Iranian revolution after its design was complete.

British Rail's consultancy arm, Transmark, had been commissioned to plan and design the upgrading of the line late in 1975. In turn, Transmark appointed Mott, Hay & Anderson as civil engineering consultant.

The 740-km route, originally built between World Wars I and II, forms part of Iran's 3,500-km network of railways and crosses Iran, connecting the Turkish and Russian frontiers. The plan was to replace the single-track, diesel operated line with a twin-track electrified service. Journey times for passenger services would be cut from 16 hours to five. The first 440 km from Tehran westwards to Mianeh was mostly to follow the existing line, with the existing embankment to be widened for about two-thirds of this distance to allow construction of the second track. The other third of this initial length was to be realigned to improve running speeds.

From Mianeh a completely new high-speed passenger link through mountainous country was planned, which would cut over 100 km off the total length of the route. Design speed was to be 160 km/h, and the project would require extensive earthworks, 14 km of tunnels and 25 km of viaducts.

Freight would continue to use the original route, which was to be upgraded to increase its capacity from 15,000 t to 100,000 t daily. The line would have become a key freight route between the Balkans and Russia.

Throughout its length the planned line would have run through difficult terrain. The whole area was highly seismic, requiring state-of-the-art design work to make the structures resistant to ground shaking and liquefaction. The mountain tunnels would have passed through complex geology, a mixture of hard and soft rock, some sedimentary and some volcanically altered rock such as schists. A combination of rock-boring machines and blasting would have been needed. To complicate matters further, the tunnels lay in an active fault zone.

The whole route was also subject to flash flooding, and Sir M. MacDonald & Partners provided a hydrologist and hydraulic specialist to assist in designing cross-drainage structures. Substantial structures to allow flash floods to escape from one side of the track to the other were needed.

Work on the design was completed just before the revolution. Although priorities have changed, the Iranian authorities are understood to be still interested in the possibility of implementing the project.

Water supply

Sir Alexander Gibb & Partners became involved in 1946 with the question of water-supply for Tehran. 'It became obvious that as part of post-war reconstruction Tehran's water supply would need urgent attention,' says former Gibb chairman John Bowcock.

Tehran at the time had no piped water-supply. The city is sited at a height of 1,500 m in the Alborz mountains on land which slopes from north to south ranging from 1900–1100 m above sea level. Traditionally, its water needs had been met by *qanats* – adits driven into the hillside until they reached aquifers which were recharged by snow from the mountain tops, and by a channel from the River Karaj.

The water was conveyed in *jubes*, or open ditches which run alongside the streets, taking advantage of the slope of the ground to distribute the water through the city. Naturally, the further it goes, the more polluted the water becomes.

In 1946 Gibb began work for the Tehran Water Board on developing the water-supply system. Gibb worked continuously on the water-supply until 1983, when a population of 7.5 million was being planned for. The firm was the only British consultant to remain in the country during the Mossadegh era in the early 1950s, when all other British companies were ordered to leave. Even when emotions were at their height, with angry crowds demonstrating in the streets against British interests, 'not one of our engineers was harmed', says Gibb technical director John Allen.

Initially, in the late 1940s, supplies were met by pumping groundwater. But Tehran's population grew at a remarkable rate, from 900,000 immediately after the war to around six million now. Since then a series of dams has been built. The first major dam was Karaj, to the west, designed by the US consultant Harza. In the late 1940s Gibb produced a report on providing water from the proposed Lar and Latiyan dams. Latiyan is on the Jaje Rud River, which flows south, and Lar is on the Lar River which flows east and then north, changing its name to Haraz and then to the Caspian Sea. A mountain range separates the two, catchments. *En route* to the Caspian, the Haraz crosses the Mazandaran plain where, with the Babol and Talar rivers, it is used in a highly developed irrigation system.

The report's initial proposal was to build Lar first and Latiyan later, but after some consideration and refinement of the report, it was decided to build Latiyan first. This was later renamed the Farahnaz Pahlavi dam after the shah's daughter, but was changed back to Latiyan after the revolution.

The contract for construction of the dam was awarded in 1960 to a consortium of three French companies – Entreprises Campenon-Bernard, Etablissements Billiard and Société Française de Dragages et de Travaux Publics. Professor Alfred Stucky of Lausanne acted as adviser to the contractor. Main construction got under way in 1960, inaugurated by the shah who placed the first bucket of concrete, and was completed in March 1967.

Construction of Lar began in 1974 to augment the supply further. It was intended that Lar water would be diverted through tunnels to the Latiyan reservoir.

Work progressed on providing the city with a water distribution system. Pipes of up to 2-m diameter were used. Polyvinyl chloride (PVC) pipes of up to 75-mm diameter were used for house connections. Others were ductile iron pipes of up to 900-mm diameter, steel up to 1,200 mm, prestressed concrete ones of up to 1,850 mm and reinforced concrete up to 2,000 mm. Gibb assisted the Tehran Water Board in establishing a pipemaking factory for the reinforced and prestressed concrete pipes. Pipelines from the Karaj River were constructed during the 1950s and 1960s to feed two water treatment plants in the city. Two further plants were built later to treat water from the Latiyan and Lar dams.

Numerous service reservoirs were also built with capacities of up to 60,000 m^3. A number of pumping stations were needed to cope with a difference in supply levels of 800 m.

Rendel Palmer & Tritton's Iranian joint venture, Irendco, often in association with John Taylor & Sons, was responsible for several schemes in provincial towns and cities. These included extending Khorramshahr's water treatment works to double throughput of 52,000 m³/d, a scheme which was completed in 1977. During the same period Irendco prepared feasibility studies, detailed designs and contract documents for extending the distribution network, adding more storage reservoirs, pumping stations and treatment works to treble the town's supply to 100,000 m³/d. Emergency measures to provide an extra 10,000 m³/day were also developed.

Latiyan and Lar dams
The Latiyan dam is 25 km north-east of Tehran and has a gross capacity of 95 million m³. Its primary purpose is to supply 80 million m³ of water annually to Tehran, but in addition 160 million m³ is supplied to irrigated areas on the Varamin plain. There is also a small hydroelectric power station, initially containing a 22.5 MW generating set but with provision to add another. The UK consulting engineer Merz and McLellan was the designer.

The hydroelectric plant is required to produce electricity only at peak times, for about two hours during the day, so the design is such that a full day's release of water passes through the turbines during the peak period. The released water is then stored in a 400,000 m³ pond behind a regulating dam about 1 km downstream, where it is drawn off daily to serve the more uniform demands of Tehran and the irrigation scheme. The regulating dam is a composite structure 45 m high, consisting of a concrete section 88 m long and a rockfill embankment section 65 m long.

The main dam is a concrete buttress structure 107 m high, with 22 buttresses at 14 m centres, and mass concrete flanks. On the left bank the dam terminates in a bored pile cut-off curtain. The ground investigation prior to construction included 110 boreholes and 915 m of exploratory tunnels. Because the dam is in an earthquake area, resistivity and seismicity surveys were undertaken, together with in-situ tests to determine the deformability and sheer strength of the rock.[1]

To improve the strength of the foundations on the left bank, where a narrow ridge of red quartzite meets Oligocene Green Beds, composed of cherts interbedded with tuffs and shales, extensive consolidation grouting was carried out in addition to providing the usual grout curtain against seepage. On the right bank, the foundation consists of a layer of poor rock contained within layers of better rock.

The downstream toe of some buttresses overlay the poor rock and mass concrete thrust blocks were built out from the toes to reach the solid red quartzite which was capable of resisting the thrust.

The foundations are drained by over 3,000m of 75-mm holes drilled from deep drainage tunnels 2.7 m in diameter. Water is collected in sumps and discharged into the regulation reservoir by automatic submersible pumps.

Earthquakes are common in the area and model tests were carried out to study the behaviour of the dam in seismic conditions. They revealed that the natural frequency of the higher buttresses lay just inside the high-energy frequencies which might be encountered in an earthquake, so the webs of the tall central buttresses were stiffened by widening them out at their downstream toes to abut one another. Both the main dam and regulating dam have low-level spillways to discharge silt washed down during floods. Spillways are designed to cope with a 1,750m^3/s flood, compared with a maximum recorded 500m^3/s.

Water travels from the reservoir to water treatment works, on the eastern side of Tehran, via the 9,576-m long, 2.7-m diameter Talow tunnel. It was driven from both ends between December 1963 and June 1966. The upstream heading passed through red and white quartzite, then limestone, before reaching the Oligocene Green Beds which were highly variable in quality and needed support from steel ribs. The tunnel then crossed the Alborz fault, which was found to consist of 0.5 m of dry compacted clay, then headed into conglomerates, mudstone, siltstone, and alluvium. Excavation from the downstream portal was entirely in alluvium.

Further studies on the Lar scheme were undertaken between 1969 and 1972, and the first two contracts were let in 1974. The dam had several functions: it would regulate flow in the River Lar for irrigation; impounded water would be transferred to the Latiyan reservoir to serve Tehran; and hydroelectric stations were included to take advantage of the 930-m difference in height between them. The dam was intended to allow Tehran's water demands to be met up to the late 1980s. The population was expected to reach 7.5 million by the early 1990s with an annual water demand of 700 million m^3, almost double the 1972 level. In addition to the water supplied from the Lar, further development of groundwater resources, and the use of treated sewage effluent for recharge and irrigation were envisaged to meet this demand.

In addition, on the Mazandaran plain a series of weirs and canals was planned to integrate the Haraz, Talar and Babol rivers in order to meet irrigation demands more efficiently. This, together with over-year flood storage, would enable the diversion of 178 million m³ of Lar water to Latiyan in an average year.[2]

The Lar dam is an embankment structure 105 m high with a crest length of 1.3 km. It was intended to impound 960 million m³ of water in a reservoir 14 km long. Construction involved 38 million m³ of cut and fill.

Water is transferred to Latiyan by a two-stage diversion scheme. A 20-km long, 3-m diameter concrete-lined pressure tunnel terminates at the Kalan hydroelectric power station, from where flow continues along a 6-km aqueduct and tunnel to discharge into the Lavarak River 3 km upstream of the Latiyan reservoir. A further, underground, power station with two 24 MW Francis turbines was constructed to exploit the 320-m head between Kalan and the river outfall.

Gibb had undertaken geotechnical investigations along 3 km of the Lar gorge to identify the best site for the dam. The geology of the area is complex but there was a recent history of volcanic activity, and it is thought that there was at one time a natural dam, formed by lava, roughly on the Lar site. The Alborz fault runs east–west along the edge of the Alborz mountain range. An earthquake of magnitude greater than 4 on the Richter scale can be expected within 10 km of the dam once every thousand years.

Ground conditions at the dam site itself consisted of lava overlain by various sediments, in particular a fine sand which was found to be liable to liquefaction in an earthquake. Extensive excavation was therefore needed to found the dam on lava.

Diaphragm walls and a grout curtain along the line of the core were provided to control seepage. Diaphragm walls ran through gravel deposits beneath the embankment, and through decomposed lava deposits on the left bank. Galleries were created in the underlying rock, 120 m below river level, to allow the grout curtain to be extended during reservoir filling if necessary.

A joint venture of Impregilo of Italy and Tessa of Iran won the contract to build the dam in 1974. Impregilo had wide experience of building dams in all parts of the world, including the huge Tarbela dam in Pakistan, from which a large number of personnel were transferred. However Lar presented some entirely new problems. One of them was the extremes of climate. At an elevation of 2,500 m above sea level, the site could expect 2–3 m of snow each year. Temperatures could drop as low as –30°C in winter, and rarely rose above zero between December and March. In summer temperatures could reach +30°C and more, although the ground remained wet and muddy for long periods after thawing. This meant that the embankment fill works had to be restricted to around six months of the year. The site had to close from December to March, and had to be made secure against the spring floods.

A critical operation was completion of the river bypass works before the winter of 1976-7 so that the next flood, which would come with the snow melt in May–June, could be diverted around the dam foundations. Despite a late thaw in 1976 the contractor achieved this objective. Impounding was due to start in October 1978, with overall completion due in September 1980.

Actual completion took place in August 1981, almost a year late, as a consequence of the disruption brought about by the Iranian revolution, and of the unexpectedly high rate of inflation that prevailed in Iran following the inception of the contract. Since the contract did not contain an adequate escalation clause compensating for the increase in costs, the resulting cash flow problems, were so severe that the contractor was unable to resume work at the beginning of the 1978 working season. Work was resumed in May 1978, after the client had agreed in principle to negotiate a suitable escalation clause. A formal agreement in this respect was signed in September 1978 and reconfirmed under the new regime after the revolution in August 1980.

Impounding of the reservoir commenced in May 1980, but starting from the summer of 1981 the balance between inflows and outflows led to the suspicion that the reservoir was leaking. In 1982 a panel of three experts, one from the UK and two from France, visited the site and issued a report expressing the opinion that there was a leakage taking place through the karstic channels in the limestone of the right abutment, externally to and underneath the existing grout curtain.

In 1983 the client replaced Gibb with Iranian consultants, the Lar Consulting Bureau, and maintained that Impregilo was responsible for the leakages. Impregilo rejected any such responsiblity and, in order to resolve the dispute, in May 1984 entered into negotiation according to the rules of international arbitration.

In 1989 further grouting to a depth of 150 m below the reservoir base was proposed. But it was reported by the UK journal, *New Civil Engineer* in 1990 that, according to a letter sent to it by Lar Consulting Bureau managing director Bijan Yazdani, more cavities had been discovered, that the fractured limestone was now thought to be 800 m, not

400 m, deep and that the depth to a watertight stratum was now thought to be 2,000 m. It began to appear that grouting would be impossible. The Lar Consulting Bureau began to question the siting and stability of the dam, stating that it had been built over a fault.[3]

Gibb was not involved in investigating the reasons for the leakage, and information from Iran has been scant. Gibb's John Bowcock says:

> It's a somewhat sad story. The dam was finished at about the time that the Shah was overthrown. At the time the dam was put into commission, our agreements in Iran were terminated. We knew that where it was built on limestone there was potentially a risk of leakage, and we made provision in our financing for a grouting programme. There was a provisional sum in the bill of quantities for this further grouting, to be done on an observational basis, as the lake filled.

In 1990 the Tehran Regional Water Board appointed a panel of experts to advise on methods to stop or reduce the leakages from the reservoir. The panel consisted of a US geologist, an Italian soil mechanics engineer and an Italian construction engineer.

Based on the same panel's recommendations, the Tehran Regional Water Board also appointed a French engineering group, Setec, to carry out investigations and submit preliminary proposals for the solution of the leakage problem.

At the end of 1993, after further exploratory drilling and detailed geologic and geophysical investigations, Setec issued a final report – with the full endorsement of the panel – concluding that it would be practically impossible to stop the leakages by drilling and grouting methods, and recommended partial blanketing of the reservoir, either with a geomembrane, or with a clay layer. Alternative solutions were also suggested in the report, including the creation of another reservoir through the construction of a second, smaller dam upstream of the existing dam.

Sewerage

A sewerage system for Tehran was not planned until the 1970s when Gibb and John Taylor & Sons, in association with a local consultant, undertook a 40-year master plan for wastewater and surface water for the whole of Tehran. The city was then one of the largest in the world without a piped sewerage system. Tehran's sewage had been disposed of in wells dug beneath each house, says Gibb technical director John Allen. As a result groundwater was beginning to be contaminated. Surface water simply ran downhill in open drains or *jubes* at the side of the street.

The master plan provided justification for the use of separated storm and foul water systems, and included social benefit/cost analyses, suggested phasing, and estimated charges to make the scheme viable. Utilisation of purified effluent and sludge in agriculture was also considered.

Gibb and Taylor were also responsible for detailed design of some parts of the sewage system including the treatment works to the west of the city. However, as far as Gibb was concerned, the amount of detailed design was limited: 'Our agreement terminated when the Shah was overthrown,' Allen says. 'We came out and haven't been back since.'

None of the master plan has yet been implemented, but recently the project has been revived. The surface water and wastewater elements of the master plan have been split, with wastewater becoming the responsibility of the Tehran Province Water & Sewerage Company, and surface water being devolved to Tehran municipality. Taylor's successor, Acer, formed by the merger of Taylor with Freeman Fox & Partners, has been advising and reviewing detailed designs by Iranian consultants on both elements.

For the wastewater scheme, now valued at £750 million, Acer's involvement has been mainly on the southern treatment plant which will ultimately serve a population of 4.8 million, and on the 20-km tunnelled main trunk sewer. World Bank funding has been sought and the project is undergoing appraisal by the Bank.

The surface water project, which includes 13.5 km of tunnel, 14 shafts including vortex drop shafts which will divert flow from existing open channels, and a 600,000 m^3 retention reservoir, has been approved for World Bank funding. It was being retendered in early 1994, with the likelihood of work starting in 1995.

Unlike sewerage systems in many other Middle East cities, Tehran's system is unlikely to suffer from pipe corrosion. Because of the sloping ground, there will be fewer pumping stations than usual, and sewage will tend to remain in the pipes for less time, providing less opportunity for septic conditions to occur. The climate is also generally more favourable there.

Taylor and Irendco also undertook sewerage schemes in the provincial towns of Rasht and Ahvaz. Rasht is a provincial capital,

350 km north of Tehran. Taylor was appointed to produce a master plan and feasibility study to serve a population projected to grow to 300,000 by 1991, and subsequently went on to do detailed design and prepare contract documents. The work was divided up, with Taylor working on the process design and detailed design for the treatment works – a conventional activated-sludge design allowing for the possibility of adding tertiary treatment later; Irendco designed the sewerage system and pumping stations.

At Ahvaz, Irendco was responsible for design and supervision of construction of stage one, from 1976 to 1978. The scheme included 190 km of sewers, two major pumping stations and a treatment works for a population of 250,000. Taylor was subconsultant for the detailed design of some of the sewers and a treatment works for a population of 22,000. Irendco also designed stage two, to cope with a further 125,000 people; this had reached contract document stage by the time of the revolution, when the project was held in abeyance.

The Azadi monument

Ove Arup & Partners was responsible for structural design of the Azadi ('freedom') monument, a 45-m arch situated at the entrance to Tehran airport at the end of a main east–west artery through Tehran. Originally known as the Shahyad Ariamehr monument, it was conceived to celebrate 2,500 years since the foundation of the Iranian empire.

Designed by a leading Iranian architect, Hossein Amanat, in a competition, it has a shape inspired by the past but modern construction techniques were used to achieve it. Indeed the introduction of new techniques was seen as a secondary purpose of building the monument.

The highly complex geometry was developed through close collaboration between the architect and engineer using computer-aided drafting techniques. The monument's main *raison d'être* is as a sculptural form, and its external appearance is therefore of prime importance. It is clad in marble, which also acts as permanent formwork for the reinforced concrete structure. Internally, there are four floors, with a 10-m high dome near the top, whose geometry is based on traditional Iranian architecture. Entrance is gained by a tunnel, to avoid door openings which would interfere with the monument's surface; this also provides access to an underground museum and library.

Ports and oil terminals

Between 1950 and 1958, Rendel Palmer & Tritton designed six T-head oil jetties at the port of Bandar Mashur. The first four could accommodate 30,000-dwt tankers. Jetties 5 and 6 had two decks and were big enough for 45,000-dwt tankers.

In the early 1960s, Rendel Palmer & Tritton was responsible for the Abadan–Tehran oil pipeline, a £25 million project for the National Iranian Oil Company. With extensions to the Caspian Sea, Meshed and Isfahan, the length totalled 2,400 km, with 16 pumping stations, tankage, and other facilities. The pipeline rises from sea level at Abadan to 2,220 m on the way to Tehran and to 1,550 m between Abadan and the Caspian.

Following a hiatus during the 1960s, Rendel Palmer & Tritton returned to Iran in the 1970s, where its joint venture Irendco won several important road commissions.

In the 1970s, both Ove Arup & Partners and Parsons, Brown & Newton were involved in a proposed commercial port development at Chahbahar (now Bandar Beheshti). Near the border between Iran and Pakistan, it was intended to stimulate the economy of the eastern provinces of Iran. The development was to have included a petroleum pier protected by a 2.15-km breakwater, two general cargo berths, a container berth, roll-on, roll-off facilities, a coastal harbour basin and a fishing harbour basin. A free trade complex within the commercial area was also planned.

Arup was responsible for detailed planning of infrastructure and roads within the port, together with administration buildings, hotels and hostels, workshops, power and water services. Parsons, Brown & Newton was responsible for the port itself. Much of the design had been completed before work was halted in 1979 by the revolution, and only the breakwater has been built.

Roads

British contractor Marples Ridgway earned the distinction of being awarded perhaps the most ill-fated road contract in Iran. The 300-km Shurgaz to Mirjawa road was to complete a link between Europe and Pakistan. Marples Ridgway won the contract in 1976 and its first problem was to mobilise £14 million worth of plant, all of which had to be imported at a time when Iran's ports were severely congested. It got round the problem by introducing roll-on roll-off ferries for some of the equipment while other machinery was delivered by road or rail. A year into the contract the contractor was perhaps fortunate to be only three months behind, with some of the plant still in customs awaiting clearance.

A search for suitable aggregate and water in the inhospitable desert terrain which the road traversed also presented considerable difficulties, especially since in the first instance the contractor was not allowed a licence to use explosives to win aggregate. Marples Ridgway hoped to make up the lost time and complete the contract in 1978, but by the time of the revolution the road was about three-quarters complete. The contractor withdrew all its staff except for a skeleton maintenance team because of the uncertainty.

Work stopped for some time. Then Iran announced that £7,500 million of foreign debts would not be paid. A number of contracts shut down completely. *New Civil Engineer* reported in November 1979 that work on the Shurgaz road was 'cautiously going ahead as payments are received on straightforward valuations'.[4] But next April, following defaults on payment by the Iranian authorities and the withdrawal of British diplomatic staff, Marples Ridgway abandoned the contract, leaving it 75 per cent complete. An argument in the Iranian courts over whether the client or the contractor was responsible for breaking the contract followed. Marples Ridgway was left with debts of £13.5 million; this was 90 per cent covered by the UK's Export Credits Guarantee Department (ECGD). The result was to make the contract a source of friction between Britain and Iran for a considerable time afterwards.

Iran today

Following the end of the Iran–Iraq War, oil revenues began to restore Iran's economy to a more healthy state. And an 'open-door' policy raised the prospect that that there might once more be opportunities for overseas firms in Iran. The market 'looked very promising' in 1991 and 1992, according to one engineer, and there were a number of trade missions which hoped to capitalise on this. Unfortunately, concurrently with the new openness, Iran's central bank passed control of foreign exchange to the commercial banks, and control was relaxed to such an extent that a flood of western luxury goods was sucked in. By 1992 Iran found itself having difficulty meeting its debt obligations. In addition, the latest five-year plan had assumed that oil revenues would be significantly higher than they were.

As a result, most of the hoped-for projects were shelved, or at best slowed down substantially. Mott MacDonald's experience was typical. It had signed an agreement to provide assistance to Mahab Ghodss, a local consultant, on a number of projects in the water sector. However, many projects have been delayed and in the first two years of the agreement only a fraction of the anticipated work went ahead.

Projects with external financing, provided by the World Bank for example, or build-operate-transfer projects, seemed more likely to succeed. In 1993 Acer, with developer Santé Trading of Cyprus, won two contracts: for the Qeshm bridge, linking Qeshm island to the mainland, and a $30 million polymer factory. The 2.5-km Qeshm bridge will cost $100 million, with Acer as design-and-build project manager and design consultant. Construction had been due to start in 1993, and although the design work had been completed, both projects were still awaiting the go-ahead in mid-1995.

chapter three

IRAQ

Iraq's natural resources make it uniquely fortunate among Arab states: it is the only one to have plentiful supplies of both oil and water. The other oil-producing states have little water, while Egypt, for example, is well-endowed with water but has comparatively little oil.

In addition, although the population of Iraq has increased considerably since the 1950s, it is neither too large to be manageable like those of some more populous countries, nor too small and scattered to form the basis of a dynamic economy.

In principle, then, Iraq is well placed to be a leading economic force in the region. In his 1958 book *Road Through Kurdistan* (Faber & Faber, 1958) civil engineer A. M. Hamilton describes four years with the Iraq Public Works Department in the 1920s and 1930s and reflects on the new prosperity which was transforming the country. Even more important than the oil wealth which the country had acquired was the Iraqi people's discovery of 'how to work together and enjoy the fruits of their natural resources in a peaceful way . . . ' Now at last there was a 'slackening in the religious and tribal feuding between Christian and Mahommedan, between Assyrian and Kurd, Kurd and Arab', he says.[1]

But this proved a transient interlude. Instead, in recent history, the development of Iraq has continued to be dogged by conflict. There has been sporadic and inconclusive civil strife with the Kurdish population in the north. Throughout most of the 1980s the Iran–Iraq War drained the country's resources and disrupted oil production, the main source of its wealth, leaving Iraq with huge debts and derailing ambitious development plans. Finally the Gulf War devastated the country's infrastructure which it will take many years to rebuild, and exacerbated the friction between the Kurds in the north, the Sunnis centred on Baghdad and the Shiites in the south.

Iraq's water has been a priceless asset since antiquity, when the ancient Mesopotamian civilisation arose in the area between the rivers Tigris and Euphrates. The engineering arts of irrigation have been practised there since that time, the Hanging Gardens of Babylon being the most celebrated example. Oil was, of course, discovered relatively recently, and began to be exploited on a major scale only in the late 1930s and 1940s.

The modern Republic of Iraq was formed from the three Ottoman provinces of Basra, Baghdad and Mosul. These were occupied by the British in World War I, and in 1920 Iraq became a British-mandated territory under the League of Nations. A year later a nominally democratic system of government was set up under King Faisal ibn Hussein. In practice, each government minister had a British 'adviser' and power remained effectively in British hands. Though independence followed in 1932, and despite a series of *coups d'état*, a small political élite backed by Britain continued to control the country.

Britain's wish to remain in control stemmed first from Iraq's strategic position on the land route to India, and later from Britain's desire to control access to Iraq's oil. The Iraq Petroleum Company, made up of British, French and US concerns, had been granted a concession to prospect for oil in 1925. Oil was first discovered in 1927 and began to flow in substantial quantities from 1934 onwards.

Following World War II a number of British consultancy firms such as Rendel Palmer & Tritton were actively engaged on infrastructure projects. Except in the oil industry, development was limited and few people were employed in industry. What had increased, from the 1930s onwards, was the number of children in education. This led in turn to an increase in the number of people from poor backgrounds gaining relatively prestigious jobs, for example in the army, and later in the Development Board, which was responsible for most major infrastructure projects under the monarchy in the 1950s. But they had no political power. Gradually such people coalesced into a highly politicised intelligentsia, opposed to the political élite and to British influence; in 1958 a revolt, which began in the army, swept away the old guard.

Several years of uncertainty followed, with disagreements between the Arab nationalists and communists over the direction that should be taken following the revolution. The issue was settled in 1968 when a group of Baathist officers took over in another coup. The basis of Baathism is a belief in a single 'Arab nation' stretching from North Africa to Iraq itself, though the Iraqi Baath party has long been estranged from the original Baathist movement in Syria.

Under Ahmad Hasan al-Bakr, who led the coup, and Saddam Hussein, to whom al-Bakr handed power in 1979, but who was effectively in charge long before that, the Baath set about concentrating power in its own hands and suppressing opposition. Its single most popular act was the nationalisation of the Iraq Petroleum Company in 1972. This was less than a year before the dramatic escalation of oil prices in 1973, which at a stroke gave Iraq the wealth to embark on a massive infrastructure development plan.

The oil revenue windfall of the 1970s also provided Iraq with the opportunity to embark on the path to prosperity enjoyed by many of its fellow oil producers. Indeed in 1976 the country launched a hugely ambitious five-year national development plan, but the war with Iran thwarted its implementation. Iraq's political leaders have generally been Sunni Muslims, though the majority of the population are Shiite. The Shiahs were less sympathetic to the idea of Arab nationalism and throughout the 1970s there had been in Iraq a growing belief, similar to that of Ayatollah Khomeini's party in Iran, that a return to a regime based on Islamic principles was needed. The revolution in Iran in 1979 raised the possibility that Iran would encourage a Shiah uprising in Iraq.

It was not only to prevent this that Iraq launched the war on Iran: Saddam Hussein believed Iran's armed forces to be weak and disorganised; he saw the opportunity to overthrow the Iranian revolutionary leaders and increase his own status in the Arab world as well as gaining control of Iran's oil reserves.

Instead of the quick victory which Saddam had anticipated, however, the two countries became embroiled in a long and debilitating war. Over 200,000 Iraqis were killed or wounded. In terms of damage to infrastructure, Iran was the worst affected; in Iraq intense fighting was limited to border areas – Basra sustained severe damage, as did many oil installations. But most of all Iraq emerged from the war with vast reserves transformed into substantial debts. Most of its construction projects, such as the Baghdad metro, had ground to a halt during the war as funds became scarce. After the war, development was constrained by the need to reschedule debts, and the need to place orders with countries willing to provide export credit guarantees or soft loans. And before the damage could be undone came the invasion of Kuwait, followed by the devastation of the Gulf War.

The activity of British firms in Iraq naturally rose and fell with the advent of Iraq's various regimes. Until 1958 the strong British influence there provided many opportunities for UK industry. Many companies pulled out in 1958, and at no time since then has the market been an easy one to break into; however some firms, including John Taylor &

Sons, Rendel Palmer & Tritton and Sir M. MacDonald & Partners, managed to continue working through the various revolutions of the 1950s and 1960s.

Development of Iraq's basic infrastructure began during the late 1940s and early 1950s, as oil production grew. In 1951, Rendel Palmer & Tritton was commissioned to carry out a comprehensive study of Iraq's railways, and to make recommendations on economics and modernisation. The oldest of Iraq's railways dated from 1902. A 1-m gauge line from Turkey to Basra via Baghdad had been constructed by Germany; it was later upgraded to standard gauge north of Baghdad. A cause of frequent comment was the name of one of the stations, 'Ur Junction', after Ur of the Chaldees, the centre of a major early civilisation.

Rendel Palmer & Tritton's study resulted in a plan for 3,000 km of new railways. Though the east–west lines from Kirkuk to Haditha, and from Baghdad to Husaiba on the Syrian border were completed, the Iran–Iraq War prevented much of the plan from being implemented.

Throughout the 1970s Iraq made further great strides towards developing an industrial base and the associated infrastructure. The national development plan (1976–80) was seen by outsiders as a wise strategy for spending oil money and for providing direct benefit to the population. Agriculture, irrigation, health and education featured prominently in the plan. In April 1979 the State Company for Construction Contracts (SCCC) alone was reported to be undertaking projects worth £1,050 million. Though the country aimed to be technologically independent it recognised that outside expertise was needed to achieve the pace of develpment it sought. Nevertheless shortfalls in materials and labour caused problems on many contracts.

Iraq was seen as an attractive, financially stable market where prompt payment could be expected, but during the 1970s many British firms found it difficult to establish a foothold. They were not aided by a trade embargo against Britain following the expulsion of eight Iraqi diplomats from Britain in 1978 – although this could sometimes be circumvented for contracts already under way. In 1979, a *New Civil Engineer* supplement on the Middle East reported Dr Kais Salem el-Bahrani, technical director-general and vice-president of the SCCC, as criticising British firms for treating Iraq as though it were still a colony. But an Iraqi chief engineer was quoted as saying: 'We have a lot against the British Government but nothing against British contractors, or English people.'[2]

Unfortunately, with the Iranian war, Iraq's reputation for financial stability faltered. Funds became scarcer, projects were delayed, jobs came to an end with no new work to take their place, and gradually most expatriate firms who had set up in the country departed, perhaps leaving a local representative where once they had a full-scale office.

But the Gulf War caused devastation on an altogether different scale. A United Nations mission to Iraq in March 1991 reported: 'Virtually all previously viable sources of fuel and power and modern means of communication are now, essentially, defunct.' Crises loomed in food and water supply. With treatment plants unable to function because of damage and lack of power, untreated sewage from Baghdad and elsewhere was being discharged directly into rivers – rivers which were also sources of drinking-water. Provision of drinking-water in Baghdad dropped to 40 l/d per person, less than a tenth of the level before the war.

The UN mission graphically summarised the situation:

> The recent conflict has wrought near-apocalyptic results upon the economic infrastructure of what had been, until January 1991, a rather highly urbanised and mechanised society . . . Iraq has, for some time to come, been relegated to a pre-industrial age, but with all the disabilities of post-industrial dependency on an intensive use of energy and technology.[3]

Bridges

Both Rendel Palmer & Tritton and Coode & Partners were engaged in building bridges across the Tigris and elsewhere as a network of metalled roads was created.

Bridges in the capital city are naturally prestigious constructions, as well as economically important ones, and between the 1950s and 1980s nine were constructed over the Tigris in Baghdad to serve the transport needs of the growing capital.

Coode & Partners was responsible for three of these, in addition to the existing Shuhada and Ahrar bridges which were both opened in 1939 to replace floating bridges dating from earlier this century. The Sarafiya road-and-rail bridge, which replaced the existing wagon ferry, was originally designed at around the same time, but it was not built until after the war. The structure totals 2.2 km in length with its approaches. Steel through trusses form the main river spans. The

A'aimma and Jumhuriya road bridges followed between 1953 and 1957.

Rendel Palmer & Tritton's 17th July bridge, originally known as the Northgate bridge, had an unusually long gestation period of 17 years from original conception to the start of construction. During this period its alignment and the number of lanes it was to carry were changed several times, and the consultant changed the construction from steel-plate girder to precast post-tensioned concrete beams. Construction eventually began in 1974, and the bridge was opened in 1978.

Elsewhere in the country, in the late 1950s, Rendel Palmer & Tritton was responsible for a pair of bridges, one road and one rail, over the Diyala River at Baquba. Though modest in size they were strategically important. The road bridge carries the Baghdad–Tehran international route. It is a 590-ft long plate girder with a main span of 250 ft. The adjacent rail bridge carries the Baghdad–Khanaquin–Kirkuk railway. Consisting of high-tensile steel trusses supported on concrete caissons it has a main span of 250 ft.

After the quadrupling of oil prices by the Organisation of Petroleum Exporting Countries (OPEC) in 1974 set off the 1970s' construction boom in the oil-producing states, one of the first British consultants to re-establish itself in Iraq was Maunsell & Partners. Within months it succeeded in landing a major bridge project.

Iraqi-born engineer Shakir al-Kubaisi, now a Maunsell director, recalls: 'In 1974 the construction boom began in the Middle East. The firm asked me to visit the area to investigate the market. I visited a few countries, including Iraq, and that led to Maunsell's decision to open an office in Baghdad in May 1975.'

To begin with a local office was set up under the name of al-Kubaisi & Partners, with al-Kubaisi and Maunsell partner David Lee as directors. Towards the end of that year the firm was rewarded with its first project, the elegant Adhamiyah cable-stayed bridge, the 11th crossing of the Tigris in Baghdad.

Maunsell was one of a selected list of international consultants invited to bid for the design of the bridge; both technical merit and price were taken into consideration in awarding the commission. Tenderers were invited to propose the form of the bridge from a list in the terms of reference for the tender. It was at al-Kubaisi's own suggestion that 'cable-stayed' had been added to the client's list of acceptable types.

At the time, few cable-stayed bridges had been built anywhere, mainly because they are highly redundant structures and thus much more difficult to analyse than, for example, suspension bridges. Development of computer techniques has increased their popularity since the 1970s. What swung the decision in Maunsell's favour, al-Kubaisi believes, was the fact that the firm had previously completed a bridge of this type, the Batman bridge, in Australia.

The choice of the cable-stayed form was not an arbitrary decision. The master plan for Baghdad's roads determined the alignment of the bridge, on a skew and at a curve of the river. Scour around the outside of the bend was very severe, and would have made it difficult to site a pier there without expensive scour protection measures. Because of this, and because of the skew, a main span of 185 m was necessary, with a single river pier on the inside of the bend. Maunsell considered a cable-stayed structure the most attractive way of providing this.

The total length of the bridge is 370 m. All the piers are founded on caissons sunk to a rock foundation. The main pier is founded in the shallow water of the dry weather channel, and is shaped to cause minimum disturbance to the river flow both in dry weather and in flood conditions. To minimise the pier width a single-plane cable system is used. The cable array is of harp form, radiating from a slender tower within the bridge's median strip.

The deck carries dual 10.5-m wide carriageways and two 3-m wide footways. Its main structural member is a 7.5-m spine box girder in high-tensile steel plate. Deep plate girders cantilever out transversely from the box and support an outer longitudinal plate girder, to improve live load distribution. All site connections are bolted, and the deck itself is a concrete waffle slab rather than an orthotropic steel plate which would have required site welding.

Detailed design was carried out between 1977 and 1979, and the bridge was built by the Japanese contractor, the Marubeni Corporation, between 1980 and 1983, at a cost of US$36 million.

Remarkably, Baghdad's bridges escaped destruction in the Gulf War; though several were damaged, sufficient repairs to allow them to remain in use were carried out relatively quickly.

Oil terminals

Rendel Palmer & Tritton was responsible for Iraq's main deepwater offshore oil terminal at Khor-al-Amaya, 25 miles from Fao and 18 miles offshore. Situated in 75 ft of water off the entrance to the Shatt al-Arab, it has a central platform measuring 200 ft by 400 ft, providing two berths for tankers of up to 100,000 dwt. Oil is pumped from the shore via two underwater pipelines. On this project, De Long pontoons on jack-up platforms were used in the construction of an offshore oil terminal for the first time. The platforms were built on the Tyne in north-east Britain and towed out to site, where their jacks were lowered to the seabed. After fulfilling their initial use of providing a working area during construction, they were eventually incorporated in the final structure as a tug berth and an accommodation platform.

The terminal was completed in 1962 at a cost of £7 million (at 1962 prices). Facilities included an office building, control tower, salt- and fresh-water supplies, sewerage, fire-fighting equipment, electrical equipment, cathodic protection and radio communication with the shore.

The Wadi Tharthar project

Historically Baghdad had been threatened, and frequently devastated, by floods of the River Tigris. These floods also laid waste valuable agricultural land downstream. During the 1950s Coode & Partners reported on numerous flood relief and water storage schemes throughout Iraq. The priority, and the first to be put into effect, was the Wadi Tharthar project. This was designed to protect Iraq by diverting Tigris flood waters into Wadi Tharthar, a large natural depression. Here the flood water could spread out to such an extent that it would all evaporate during the summer.

The scheme comprises a 225-m long barrage, with 17 gates, across the Tigris 60 km north-west of Baghdad. A 500-m long escape regulator with 36 gates controls the release of excess flood waters into the 53-km long escape channel which carries the water to the Tharthar depression. Construction required removal of 40 million m^3 of material. The barrage and escape regulator carry the main highway linking Baghdad with the north. The main Baghdad–Mosul railway also crosses the regulator.

Although the primary purpose of the scheme was flood relief, the barrage also included headworks for a projected irrigation canal and an intake for a low head hydroelectric power station.

Sewerage

Baghdad already had an extensive water-supply system which had been constructed in stages following the creation of the Baghdad Water Board in 1924. By 1946 a water treatment works with a total capacity of 15 million g/d had been installed. But although parts of the city had a stormwater drainage system, there was almost no foul sewerage before the early 1960s. The city's experience followed that of many other Middle Eastern centres of population in that, once they had been provided with a water-supply system, the need for a sewage system became apparent.

The population was increasing rapidly – between 1945 and 1965 the city's population rose from 500,000 to 1.75 million and was projected to reach 3 million by 1990. With the completion of the Wadi Tharthar project, the settled area was also increasing, with development spilling out from within the bunds which protected the old city limits onto the surrounding plain. In 1947, the British firm John Taylor & Sons was commissioned to report on the provision of sewerage for Baghdad.

John Taylor & Sons is said to have been recommended for the job by one of Coode's partners who was at the time working on bridge and irrigation schemes and was asked by the Mayor of Baghdad if he knew of a consultant who could do a sewerage scheme for the capital. He immediately thought of Godfrey Taylor, whom he knew well through their membership of St Stephen's Club.

Godfrey Taylor, senior partner at the time, visited Baghdad in 1948 to do a survey and report for a fee of 1,000 guineas (he was paid £1,000 and spent the next few months trying to recover the £50 shortfall).

His report set out the broad principles of the scheme which was eventually built, but it was eight years before the firm received instructions to begin detailed survey and design work. Two years before this a French consultant had done another report free of charge but John Taylor & Sons, in what was probably its first commercially motivated piece of salesmanship, responded by updating their earlier report free of charge, and as a result beat off the French challenge. John Taylor & Sons worked on the project until the early 1970s, when the work was taken over by Haiste International, under partner Bernard Eagles.

The first-stage works were to serve part of the east bank area of the city. Tenders were invited in 1957, but when only one offer was received the project was re-advertised as three separate contracts. This time the response was more satisfactory but a few days after the tender closing date the revolution took place and the monarchy was overthrown. This not surprisingly prevented the contracts being awarded.

Although the new regime decided to proceed with the scheme, it withdrew the site originally proposed at Soiada for the main treatment works. The only alternative was 7 km away on the Diyala River. Major new surveys were therefore needed as well as modifications to the design of the works and the trunk sewer, and negotiation of variations with the contractors. This was nevertheless completed in time for work to begin on all contracts in the autumn of 1959.

In the initial stages, there was no reliable information on probable flows on which to base the design. However, a decision to adopt 115 l/d per head as the dry weather flow and 450 l/d per head as the peak flow proved accurate, and subsequent extensions used the same criteria. Good quality Iraqi cement was available. Precast concrete pipes were used for the smaller sewers, in-situ reinforced concrete for the larger trunk sewers, and locally made asbestos cement pipes for most of the rising mains and some of the house laterals. Sulphate-resisting cement was used throughout. The minimum sewer diameter was 225 mm because it was considered that the risk of blockage was unusually high.

Concrete pipes were first laid with cement-mortar joints as was standard British practice at the time, but it was found that the joints cracked because of the extremely low humidity and wide temperature variations. As a result, John Taylor & Sons and the German contractor Strabag developed a more flexible bitumen-cement jointing system. A hot mixture of cement and bitumen was poured into the annular space between adjacent pipe sections, the liquid being retained by a rubber ring clamped around the pipe barrel. After the bitumen/cement mixture had cooled the rubber ring was removed and a fillet of cement placed round the end of the pipe. This method produced joints of sufficient flexibility.

Another problem that mystified the engineers for some time was that concrete pipes cast in the afternoon leaked when tested, while those in the morning did not. After considerable investigation it was found that workers had their lunch around the pile of aggregate and spat out their date stones into it. These stones were the source of the leaks.

Because the ground in Baghdad was so flat, many pumping stations were needed – over 30 on the east bank alone. A small number of standardised designs were adopted. Rising mains were kept as short as possible because of the risk of corrosion caused by the potential for septic conditions.

Progressive development of the sewerage system followed with a secondary treatment plant and extensions to the system being commissioned in 1966 and a further extension to the northern suburbs and into the Karradah peninsula in 1970.

The treatment works was sited at Rustamiyah, 7 km from Baghdad on the banks of the River Diyala. It was essentially a conventional activated-sludge plant working on the diffused air principle. Under normal conditions the works was designed to produce effluent to the UK Royal Commission standard which at that time formed the basis for British practice. This called for a five-day biochemical oxygen demand of 20 parts per million (ppm), and a suspended solids content of 30 ppm from a population of 300,000.

Because of the infrequent but heavy rainfall, flows over twice dry weather flow (DWF) were diverted directly to the outfall via a bypass. The whole site was surrounded by a bund to protect it against flooding. Power was initially generated on site by three 620-hp alternators. In the early 1970s, after the construction of new power stations in Baghdad, the works was connected to the grid at the same time as it was extended to cope with a population of 750,000.

It was in Baghdad that a serious problem for engineers in the Gulf first became evident: that of corrosion of concrete sewers. It first manifested itself in the in-situ trunk sewer leading to the treatment works, and in the inspection chambers into which the asbestos cement pumping mains discharged.

The problem was hydrogen sulphide attack. *New Civil Engineer* said in February 1976: 'The Middle East has all the classic conditions needed for the development of rapid attack.' Low flows in the sewers and the flat land meant that sewage remained in the pumping mains for long periods, allowing septic conditions to develop. Hydrogen sulphide is produced which, with the aid of bacteria, oxidises to sulphuric acid which attacks the concrete. High temperatures help to speed up the process.[4]

The phenomenon was not unknown and precautions had been taken in the sewer designs. But the speed of the attack came as a surprise.

On Baghdad's east bank John Taylor & Sons with Sika introduced a major programme to apply a bituminous lining inside the large diameter in situ trunk sewers. This was never wholly satisfactory because external groundwater percolated through the concrete as vapour and prevented complete adhesion between the lining and the concrete.

Coal-tar epoxies manufactured by Sika were later tried and proved more successful.

For precast pipes of about 1 m diameter a technique was developed whereby a plastic former was threaded inside a length of pipe which had been laid, then the annular space between was grouted up. Epoxy internal coatings to pipes already laid also proved successful. A similar technique was developed for new pipes, whereby the concrete was cast around a mechanically keyed PVC liner.

Later, in 1977, a further study of the sewer corrosion was made by another consultant, Haiste International. It recommended the use of a preformed factory-made glass reinforced plastic pipe liner as the most effective long term remedy to the corrosion problem. This method was used on badly corroded sections of the main trunk sewer which had not previously been treated. The liners were manufactured in a new factory which had been set up in Dubai.

This was an innovative technique at the time, and helped to establish a relationship with the Sewerage Board which stood the consultant in good stead when a contract for a new sewerage master plan came to be awarded. The study envisaged a population of 4.3 million through to the year 2000.

Haiste International senior engineer Hugh Allan recalls:

> We started the master plan study in July 1980, and were well into it by September, when the Iran–Iraq War started. We were stuck in Baghdad with no way out, and once over the initial trauma of being under fire the one thing we could do to keep our minds off the potential worst scenarios was to get on with the job. The aerial bombardment soon ceased, apart from occasional forays, when the Iranian air force ran out of spares for their Phantoms . . .

Other colleagues found safe refuge during bombing raids in the trunk sewers under construction.

A decade later history repeated itself when Haiste staff working on Basrah sewage treatment works, near the border with Kuwait, were among expatriates held hostage as part of Saddam Hussein's 'human shield'. Though they were eventually released this underlines the peril which working in the region could present.

Following on from the 1980 master plan study, the firm undertook detail design of new sewers, pumping stations and extended treatment plants to serve the future development areas of Baghdad. At that time the city was growing at a rate of around 5 per cent yearly. Allan says:

> In order to acclerate the construction programme and meet the government's ambitious target, the Sewerage Board let a number of huge design and construct sewerage contracts concurrently with our master plan study, and appointed two Polish engineers on a secondment basis to assist them with this mammoth task. It quickly became apparent that they couldn't hope to control such projects with their existing resources. As a result we were awarded a further commission to undertake design approvals and advise on construction supervision and control. At peak the firm had more than 50 expatriate staff in Baghdad.

Some of the contracts for the future areas were also implemented. The total expenditure at 1980 values amounted to $1,000 million. These works included a combined foul and storm drainage pumping station, built as a design and construct contract by Japanese contractor Marubeni and believed to be the largest in the region, with a capacity of 45 m^3 / sec

The impetus behind this huge level of activity was partly the result of the availability of increased oil revenues, but mainly due to the then new president Saddam Hussein. Allan remembers:

> He was determined to build a modern Baghdad which could stand comparison with any other Arab capital city and he wanted to do it within an incredibly short time span. He was an extremely powerful and ruthless man even then, and you can imagine that the pressure to get things done was felt by our client and passed on down the line – both to us and the contractors who carried out the work, often in chaotic conditions.

In what was a novel commission for a consultant at the time, Haiste International also provided operation and maintenance training for the Sewerage Board staff who would have responsiblity for operating the new facilities. But this became progressively less viable as key Sewerage Board workers were called up for war service, and the consultant's staff ended up doing the work for themselves instead of training others, and the commission was brought to an early end.

Water supply

Demand for water grew inexorably in Iraq as society became increasingly urbanised. Binnie & Partners was engaged in two major water supply schemes in the 1970s and 1980s.

The first of these was for the Karkh area of Baghdad where in 1978 the consultant carried out a study of water-supply needs up to the year 2000. The population of Karkh at the time was 1.5 million, projected to grow to 2.5 million. The firm recommended that its proposed scheme should be constructed in a single turnkey contract. It included an intake on the Tigris 40 km upstream of the city, with a treatment works with a capacity of 1,365 million l/d nearby. A total of 40 km of dual 2.1-m and 2.3-m pipelines carried the water to four reinforced concrete service reservoirs with capacities ranging from 28 million to 215 million litres. There were 160 km of ductile iron primary distribution pipe for the 450-sq km distribution area, and 1,425 km of secondary and tertiary pipeline. The project was valued at £550 million.

The scheme includes a complete supervisory control and data acquisition (SCADA) telemetry system linking control centres at the treatment works and six of the main pumping stations. The system came on stream between 1985 and 1987.

Binnie & Partners also designed a scheme for Basra and surrounding towns, with a total population of 900,000, in 1985. This scheme reached tender stage in 1988. Plans assumed that the population would ultimately rise to 1.75 million if a proposed new city to take overspill from Basra were built. Existing water supplies were taken from the Shatt al-Arab but this supply was becoming increasingly brackish as a result of upstream irrigation activities. Stage I of the new scheme was to provide 455 million l/d, stage II a further 682 million l/d.

Binnie investigated several possible supply options and recommended using the River Gharraf, about 200 km north-west of Basra. Other sources existed nearer to Basra but would have required 'unprecedentedly large' reverse osmosis desalination plants.

The main supply pipeline would end at a terminal reservoir where the new supply would be blended with 46 million l/d of brackish water from the Shatt al-Arab. Separate terminal reservoirs were to serve Zubair, Umm Qasr, al-Bakr and Shuaiba. Tenders were being assessed when the project was put on hold indefinitely.

Airports

Baghdad's first post-World War II international airport had been designed by Sir Alexander Gibb & Partners in the 1950s and built over the following decade, but by the 1970s it was no longer adequate. It was decided to build a completely new airport to be known as Saddam International Airport. Full tender design of the US$1,000 million project was another early job for Maunsell after the firm re-established a presence in Iraq; it won the job with the proviso that it must complete the tender design in an almost impossibly short time of six months. The firm rose to the challenge: holidays were cancelled, staff worked a seven-day week. 'We had 60 per cent of the London office staff, plus the subconsultants, working on it – 225 people in all,' says al-Kubaisi. 'Even the senior partner, John Baxter, had a drawing-board in his office, where he worked on the layout of the fuel farm.'

The scheme allowed for phased development to cater for anticipated traffic up to the year 2000, with all the usual facilities of a modern international airport. This included a 4,000-m runway, aprons for 25 aircraft, a 12,000-m^2 cargo handling building, a control tower and communications building, as well as water-supply, sewerage and drainage to serve a population of 20,000. On the completion of Saddam Airport, Maunsell was commissioned to do the tender designs for Basrah International Airport, which was completed in a very short time. Maunsell had previously prepared the master plan for the airport.

Irrigation

As Iraq's development and industrialisation gathered pace, people migrated from the countryside to the towns. Food production gradually declined and the country became a net importer of food. This was one reason for interest in large-scale irrigation schemes.

Sir M. MacDonald & Partners designed and supervised several large irrigation projects from the 1960s to the 1980s. Typical of these was the Lower Khalis irrigation scheme commissioned by the Ministry for Agriculture & Agrarian Reform. This covered a 93,000-ha area between the Tigris and the Diyala. Historically it had been irrigated, and to a certain extent still was, though areas of soil had become saline through over-irrigation. However, the idea behind the Lower Khalis scheme was to use the water of the Diyala to crop the whole area between the two rivers. Drainage as well as irrigation would be provided and the saline soil would be reclaimed.

In the mid-to-late 1960s, the firm had also been responsible for the Diyala weir and Upper Khalis system which irrigated around 60,000 ha. The weir is a 470-m long concrete gravity structure designed to be able to pass safely a 1-in-100-year flood flow of 4,000 m³/s. The structure incorporates 23 12-m wide weir crest gates, canal headworks with eight 8-m wide scour sluice gates, seven 8-m wide canal head gates, and a prestressed concrete road bridge. Flow is diverted to feed irrigation canals on the right and left banks. The Upper Khalis scheme, with almost 60 km of main canals, was built immediately beyond the weir.

Construction of the Lower Khalis scheme followed a separate feasibility study. Detailed design began in 1972 and construction followed between 1976 and 1984. A total of 55 km of main canals, 540 km of branch and distributary canals and 1,200 km of watercourses, all concrete-lined, distributed the water. The basic unit of land was the 'watercourse unit', a rectangular 60-ha area of land fed by a single offtake channel through the middle. These were further subdivided into 7.5-ha farm units into which watercourses were ploughed. Each farm unit was levelled to a horizontal plane with an accuracy of +/– 25 mm using lasers, with a system originally developed for land reclamation in Iraq by Spectra-Physics in 1975. For Lower Khalis a fully automated version was introduced. A laser transmitter generates a laser plane over the unit. Sensors on board the earthmoving plant detect the laser and the position of the machine's scraper blade in relation to the required level, and adjust the height of the blade automatically. Alternatively, the earthmover's driver can be provided with a visual display indicating the blade position, allowing the driver to adjust the blade position manually.

Iraq's low-permeability soil is suited to surface irrigation. Thus, once the farm units had been accurately levelled, water could be introduced at the top of the unit and allowed to flow across it. This replaced the traditional system of surrounding each unit with earth embankments.

The trapezoidal concrete canals were slipformed using a concrete paving machine. Most could be paved as a single unit, says Sir M. MacDonald's Jim Perry, deputy resident engineer on the project, but the 51-m³/s main channel, with a 4.5-m bed width, had to be built in two passes.

Another of the scheme's innovative features was its highly sophisticated remote control and monitoring system. This used power-operated radial gates and moveable weirs to control flows throughout the system. French-made Neyrtec distributors were used which gave a constant discharge with a fluctuating upstream head, ensuring that each farmer received a fixed quantity of water. Data from canal regulators, tail escapes and pump stations were transmitted to a central computer system which displayed in a semi-graphical form information such as water-levels, gate positions and discharges. All gate positions could be adjusted from the central control point. The control system was designed by Mott, Hay & Anderson and was installed by Japanese contractor Sumitomo Corporation.

The scheme also dealt with the high levels of soil salinity in some areas. Soil salinity is a hazard inherent in traditional flooding methods of irrigation where no drainage system is provided to complement the irrigation system. The irrigation water causes a gradual rise in the groundwater table until it is close enough for water to rise to the ground surface through capillary attraction. As it evaporates, an increasing amount of salt is left behind which prevents any crops being grown. A white film of salt covers the ground.

In Lower Khalis the soil was reclaimed by ponding water on the saline areas, allowing it to leach out the salt before draining away. The existence of a drainage system prevents the water-table from rising so that the phenomenon cannot recur.

Access to all parts of an irrigation system is obviously vital, and a 300-km network of surfaced all-weather roads formed part of the project. They linked the villages throughout the project area with each other and with the national highway network. The roads, 4-m or 6-m wide, were built with a crushed stone sub-base, a cement-bound granular road base, and a bituminous surface dressing. Raised earth roads linked to this feeder road system allowed access to all fields. Other infrastructure work included a building complex with 270 houses, a headquarters office, a telephone network, power supply and sewerage. An agricultural experimental station was also built, to allow the most suitable practices for agriculture, horticulture, livestock husbandry and farm management to be developed.

Sir M. MacDonald & Partners was also responsible for the Dalmaj irrigation project, fed from the Kut barrage on the Tigris and built due east at around the same time, and for the East Gharraf project. Other schemes, such as the Greater Mussayib project (to rehabilitate an existing system), were designed but did not go ahead as funds became scarce because of the Iran–Iraq War.

Baghdad Metro

Perhaps the biggest single project to be scrapped because of the war was the Baghdad metro system, valued at £5,500 million in the mid-1980s. This would have vied with Cairo's metro (whose first stage was commissioned in the late 1980s) for the position of the first rapid transit system in the Middle East. It was to be built as part of a balanced system covering improvements to all means of transport in Baghdad. The initial phase was to have been 32-km long with 36 stations. A consortium of 10 British consultants – W. S. Atkins, Design Research Unit, Freeman Fox & Partners, Sir William Halcrow & Partners, Halcrow Fox, Charles Haswell & Partners, Henderson Busby, Kennedy & Donkin, Merz and McLellan and Rendel Palmer & Tritton – collectively known as British Metro Consultants Group – was commissioned to design the project.

'Design began in 1980, and at one time 100 expatriates and 50 families, were based at the design office near Baghdad airport. Hostilities made this an uncomfortable time, as Scud missiles were landing on the city,' says Viv Hoad, who was seconded from Halcrow to be British executive director on the project from 1984–6. 'The airport was said to be at risk,' continues Hoad. 'It was difficult to know what to do. In the end we plotted on a map where the missiles landed and found that all had fallen within 2.5 km of the centre of Baghdad – so we reckoned we were far enough away to be safe.'

The design was completed, but the consortium's 10-year appointment ran out in 1989, when construction had originally been intended to end, without a start having been made on it.

Aftermath of the war

Iraq's debts at the end of the war with Iran meant that rebuilding work depended on outside loans and export credits. Because the UK was unwilling to give export credits to what was now seen as a risky country, British firms did not become heavily involved in reconstruction projects. Iraq had barely begun to find its feet again before much of its infrastructure was comprehensively destroyed in the 1990–1 Gulf War. After the war with Iran, some construction firms had had the opportunity to rehabilitate projects they originally designed or built. They are unlikely to get that chance again, at least while the political climate remains unsettled.

Working in the Middle East could bring unexpected hazards. On p.212 Viv Hoad describes how he was caught up in the revolution in Libya. Cecil Rogers of Gibb found himself in a personal predicament which must have been no less worrying. He had been working for five years as resident engineer on Baghdad's first international airport when in 1970 he found himself unable to leave the country and virtually under house arrest.

'The airport was built in two stages,' he recalls. 'Stage one was started in 1958 and finished in 1965, and this included all the ground works. But the Iraqis wouldn't take it over until it was finished completely.

'By the time it came to be commissioned in 1970 the cast iron mains had been in the ground for over ten years. The electrical services had never been powered.' The result was that there were a lot of teething troubles with the services. Rogers was refused permission to leave the country.

'I was held as hostage. I was stuck out there for 10 months. They took away my passport and stationed a soldier outside my house. I was allowed to go to the Alwiyah club. But I had to report to the British Embassy every 48 hours. My secretary was in jail for six weeks.'

The rest of Gibb's 25 staff and their wives were allowed to leave. 'It was very worrying,' he says phlegmatically. 'Eventually a fellow I was very friendly with, Donald Hawley, a chargé in the embassy, got me out. The Iraqis wanted engineers to service Hunters [British jet fighters] and he said: "no, we're not sending anybody while you're holding other British citizens".' Evidently the Iraqi Ministry of Defence was stronger than the Ministry of Works because Rogers was soon released.

The experience did not discourage him from working abroad again – though not in Iraq.

B A H R A I N

———

Bahrain, the Dilmun of 4,000 years ago, is rich in ancient historic sites as evidenced by the thousands of ancient burial mounds in the north. Situated on the maritime route between the major civilisations of Mesopotamia and the Indus, it was important as a trading centre and supplier of provisions for passing vessels, including fresh water drawn from springs under the sea.

Discovery of oil in 1932 made Bahrain the first Gulf state to acquire oil wealth. It also transformed the prospects for the whole region, by concentrating oil companies' efforts on the western shores of the Gulf, including Kuwait, and the eastern part of Saudi Arabia. Until then many geologists had believed that oil would not be discovered there in commercial quantities, because the Oligocene–Miocene strata which occurred in Iran and Iraq were not present.

In 1931 the Emir of Bahrain granted an exploration concession to Standard Oil of California, ignoring British government pressure to award the concession to a British firm. Oil was discovered in commercial quantities after just six months' drilling. The immediate result was intense interest in exploring the western shores of the Gulf – Standard Oil sought, and in 1933 won, from King ibn Saud, a concession to explore in Saudi Arabia.

Bahrain's oil reserves are nonetheless modest – they are expected to last only till 2000 – and there seems little prospect of finding more oil, despite extensive exploration. The state has therefore attempted to diversify its economy away from oil production, which has declined slowly since peaking at 76,000 b/d in 1970.

The State of Bahrain consists of 33 islands, only three of which are inhabited. Most of the population is concentrated on Bahrain island itself, in the capital, Manama. It is linked by causeway to the second largest island, Muharraq, which is the site of Bahrain's airport and which has considerable strategic importance as a stopover point. Causeways also link Manama to the third most heavily populated island, Sitra, and to Saudi Arabia. Rainfall is only 70 mm a year, but aquifers which carry fresh water under the Gulf from Saudi Arabia provide groundwater and have made Bahrain one of the most fertile areas of the Gulf.

Historically Islam has been the religion of Bahrain since shortly after the death of the Prophet Muhammad. Bahrain was strongly influenced by Iran and from 1603 to 1783 came under Iranian rule. The island's governor was overthrown in that year by the al-Khalifah family, who have retained power ever since.

Early last century Britain became increasingly concerned about the Gulf, partly because of acts of piracy there and partly because of its strategic position on the trade route to India. From 1816 onwards a series of treaties was signed between Britain and Bahrain. The first was a 'treaty of friendship' but a second peace treaty was needed in 1820 before piracy was ended. Further treaties in 1880 and 1892 turned the state effectively into a British protectorate, denying it the right to an independent foreign policy.

Finally, in 1923, the existing ruler was deposed and a British 'adviser' appointed to assist the new ruler, a situation which continued till 1971 when Britain left Bahrain. The advent of oil wealth in the 1930s led to improved literacy and the creation of an industrial working class. Periodic unrest characterised this period, with demands for a legislative assembly and trade union rights. There were demonstrations over the Suez crisis in 1956 and strikes, including one in the oil industry which escalated into a general strike in 1965. In general, political dissent was repressed, but Bahrain's reputation for radicalism was a factor in its exclusion from the United Arab Emirates.

When Britain decided to pull out in 1970, a new constitution was drawn up which allowed for limited popular elections to a legislative assembly, though this was dissolved in 1975. Following the Iranian revolution in 1979 tension arose between the majority of Bahrain's population, who are Shiah Muslims, and its rulers, who are Sunni. This led to the signing of a formal defence agreement with Saudi Arabia in 1981.

Economically, before the discovery of oil, Bahrain was one of the most prosperous places in the Gulf, with industries such as pearl-diving, boatbuilding, and fishing. It was also an important trading entrepôt. The basis of modern Bahrain's prosperity, however, was originally oil. But with production declining, Bahrain has had considerable success in diversifying its economy and in the early 1980s oil contributed only 55 per cent of government revenue. Oil refining at Sitra, where crude oil, mainly from Saudi Arabia, is processed, continues to be important, however.

Most importantly Bahrain has developed as a telecommunications and offshore banking centre. Industrial diversification has included the building of an aluminium smelter which benefits from the availability of cheap energy supplies, and the construction of the Arab Ship Repair Yard (ASRY) and dry dock. An aluminium extrusion plant, and paint and explosives factories, have been built, and tourism is being encouraged, taking advantage of Bahrain's excellent air links.

Expansion of the airport has been one of the key factors in Bahrain's postwar development. The island had been a stopover point for flying boat services in the early days of air travel, as was Dubai in the United Arab Emirates. Travel by air before the days of jet airliners was leisurely. Flights took place in daylight only, with a series of overnight stops *en route* to destinations further east. Gulf Air, whose headquarters is in Bahrain, was originally a subsidiary of the British Overseas Airways Corporation (BOAC). BOAC and Gulf Air operated Speedbird House as a rest-house on Bahrain (it was renowned for its excellent curries).

Hotels

The strategic importance of Bahrain as an air hub was underlined in 1966 when Gulf Aviation, then an associate company of BOAC, invited tenders for a turnkey contract to build a prestigious new hotel on the outskirts of Manama. The Gulf Hotel was the first of a series of up-market hotels to be built in the region.

The UK construction firm Wimpey won the contract, which gave it responsibility for everything from the structure to furnishings in January 1967, and construction began in October that year. The site was on the foreshore between high and low water-levels to the east of the capital. Preliminary work included the creation of an 18,500 m² artificial peninsula providing space for the building, swimming-pool, approach roads and car park.

A soil investigation found stiff clay and mudstone over the site beneath a surface deposit of sand and shells. Foundations consisted of column pads placed in the clay or mudstone 2 m below seabed level within a bund of well-compacted desert fill. Because of the closeness of the site to the sea, difficulties caused by the high water-table were expected, and it was anticipated that sheet piles might be needed to line the excavations. In fact a period of low tides coincided with construction and the foundations were built using only timber supports, supplemented by pumping when necessary to keep them dry.

Structurally, the hotel has an in-situ reinforced concrete frame with columns generally on a 3.8-m x 4.6-m grid. Sulphate-resisting cement was used throughout. It is V-shaped in plan, to give each room a view of the sea. A central reinforced concrete core, at the apex of the V, contains the lift shaft and main staircase. There are 120 guest bedrooms and a further 10 for VIPs.

Reception is at first-floor level, approached by twin vehicle ramps. A lounge and the main dining-room, also with a view over the sea, are at the same level. Floors two to six house the bedrooms. The ground floor is taken up mainly by plant rooms for air-conditioning and water services, cold and general storage areas, unloading bays, staff facilities and a banqueting hall.

Finishes were to the high standard which has since become customary in hotels, with marble for the main entrance and extensive use of terrazzo tiling for stairs and balconies. Bahraini teak was also extensively used.

The hotel was opened only 700 days after the first load of desert fill was placed at the start of the land reclamation contract. Wimpey later undertook an extension contract providing additional space in both wings.

Bahrain airport

Shortly afterwards, in 1969, the Bahraini authorities began to consider expansion of the airport itself, and appointed the Athens-based consultancy Gibb Petermuller & Partners to prepare a master plan and traffic forecasts for the long-term development of Bahrain airport. Gibb Petermuller was a joint venture between the British consultant Sir Alexander Gibb & Partners and Austrian architect Alecco Petermuller. Formed in 1965 with the intention of winning work in North Africa as well as the Middle East, it met with outstanding success in the latter region.

Gibb Petermuller recommended a new and much larger terminal building, catering facilities and apron, with a view to accommodating the new long-distance jumbo jets which were then starting to be built. The recommendations were accepted, and Gibb Petermuller was further appointed for detailed design and supervision of the new facilities. Contractor John Howard began work almost at once, and the new facilities were ready for the first jumbo by October 1971.

It was a particularly timely move. Tom Weaver, Gibb Petermuller's resident partner in Bahrain, told *New Civil Engineer* in 1977:

> It coincided not just with the arrival of the jumbos, but with a worldwide increase in flights to the Far East, and also the Indo-Pakistan war which caused traffic to divert. Once diverted, much of the traffic continued to use Bahrain, making it one of the world's great air travel staging posts.[1]

The new terminal building was structurally unique in Bahrain, being made of steel. Traditionally, until the 1950s when concrete began to be used, masonry had been the main construction material in the state, but by the early 1970s many buildings were showing signs of deterioration caused either by contaminants in the aggregates, or by aggressive conditions. So while a steel frame incurred high initial costs, over its whole life it might well prove less expensive than concrete. Weaver claimed: 'Discovery of "Bahrain cancer", as the concrete corrosion was called, and the performance of the airport structure, have combined to convince other consultants that the use of steel is economically viable.'

Gibb Petermuller's involvement continued with subsequent extensions to the airport facilities, beginning with the addition of a 465-m² VIP suite between 1973 and 1975. Three lounges were provided, including the Ruler's Lounge, designed for private or official receptions by the Emir or his ministers.

In 1980 Gibb assisted British Airports International with the preparation of an updated master plan. This confirmed the validity of the initial planning and the viability of the development, and recommended considerable expansion to meet expected increases in traffic demand to the end of the century. Detailed design and supervision of construction of the master plan proposals have continued since 1982. Works include land reclamation to provide for installation of navigation aids, safety areas and space for extension of parking aprons and other facilities. Runways and taxiways were resurfaced and a new lighting system installed while maintaining 24-hour-a-day flying. A cargo terminal with a design annual throughput of 57,300 t was built between 1984 and 1987. Altogether, US$120 million was spent between 1970 and 1990.

In 1985 the government appointed another consultant, Scott Wilson Kirkpatrick, as designer and project manager to double the size of the international terminal building and refurbish the existing facilities, at a cost of £46 million. In phase I a mirror image extension of the original building was constructed, allowing for total segregation of arriving and departing passengers to improve security. Like the original building, the extension is a conventional steel-framed structure on piled foundations, with reinforced concrete floors. Externally the building is glazed and clad to match the original. A new elevated road structure in reinforced and prestressed concrete provides access at departures level. Phase I was completed in October 1991 after which refurbishment of the original building to the same standard as the new one began.

Industrial development

Gibb was appointed in 1972 to work on another extremely prestigious project in Bahrain. The Organisation of Arab Petroleum Exporting Countries commissioned the firm to carry out a feasibility study for the construction of a dry dock and ship repair yard. The following year Gibb and Profabril of Portugal were appointed jointly as consulting engineers on the US$185 million project, which was part of Bahrain's strategy to diversify its economy away from oil. Gibb Petermuller & Partners was commissioned to design the buildings and workshops for the project.

While not on the same scale as the contemporaneous Dubai Dry Dock, the Arab Ship Repair Yard was still a massive undertaking. Designed purposely for the repair of very large crude carriers (VLCCs), it occupies 50 ha of land reclaimed from the sea 7 km off the southern tip of Muharraq Island and is reached by a causeway.

Construction, by the south Korean contractor Hyundai, took just two years. The Koreans, who had recently moved into Bahrain and naturally wanted to make a good impression, quickly gained a reputation for thoroughness. They even went as far as to build their own cement factory to be sure of supplies.

Infrastructure services designed by Gibb Profabril included: 250,000 m^2 of paved roads and storage areas; an extensive system of surface and foul water sewers, sewage pumping stations and treatment plant; an electrical distribution network with one main, and 12 secondary, substations; generating stations for compressed air and acetylene and a liquid oxygen station; distribution networks for potable water; salt water for fire-fighting, and for oxygen, acetylene and compressed air; and foundations for a desalination plant and fresh water storage reservoir.

Gibb Petermuller designed 20,000 m^2 of workshops and 15,000 m^2 of administrative buildings to cater for a labour force which was planned to reach 2,000 by 1980. All the workshops were of steel construction and were designed to cut down sunlight but encourage natural ventilation. A system of controllable high-level exhaust louvres is used in conjunction with low-level intake slot vents between the cladding and internal block walls, providing natural convective air circulation throughout the workshops and reducing the air-conditioning load.

The ship repair yard enjoyed considerable success though its operations were adversely affected by the Iran–Iraq War.

Crucial to Bahrain's plans to diversify its economy away from oil was the aluminium smelter, built in 1968 and the Gulf's first heavy industrial enterprise. At first, the idea of building a smelter in Bahrain seemed bizarre since Bahrain has no bauxite, the raw material from which the metal is extracted. However, the Gulf is geographically well situated between the source of bauxite, Western Australia, and the potential markets for the finished product in Europe. But even more important are Bahrain's reserves of natural gas. Aluminium is extracted from bauxite by electrolysis, an energy-intensive process, so the availability of cheap energy makes a crucial difference to the viability of the operation.

The client for the project was Aluminium Bahrain (Alba), a consortium headed by the Bahrain government as principal shareholder with participation from Britain, Germany, Sweden and the US. The main contractor was British Smelter Constructions, a consortium comprising Wimpey, John Brown Engineering (Clydebank) and the British aluminium trading company Amari. Wimpey was responsible for design and construction of all civil and structural engineering and architectural works, as well as for some mechanical engineering aspects of the plant.

Plans for the smelter expanded during construction so that the initial contract value of £24 million grew to £54 million, resulting in a plant with capacity to produce 90,000 t of aluminium a year and which could be extended to 120,000 t.

The smelter, which came on stream in 1970, is situated on Bahrain's east coast, two miles south of the Bahrain Petroleum Company's refinery which supplies natural gas to the plant from purpose-drilled wells in the Khuff gasfield. The gas turbine power station, the biggest of its kind in the world at the time it was built, houses 14 13.025-MW, and four 16.4-MW, gas turbine units supplied and installed by John Brown. They are housed in a steel portal framed structure 152 m long.

Pre-construction work on the smelter site included excavating 54,000 m³ of limestone. Much of this was used to form a reclaimed island where bauxite would be stored. Bauxite is carried from the island to the smelter by a 9.6 km aerial ropeway. A 180 m jetty provides berthing facilities for bulk carriers of up to 35,000 dwt. The reclaimed island is situated in relatively shallow water which shelves steeply away to provide sufficient depth for the bulk carriers.

Electrolytic reduction of the raw material takes place in pots, rectangular steel shells 9 m long by 3 m wide and 1.5 m deep. Four pot rooms, 650 m long and 23 m wide, each house 114 pots. Because of the repetition involved in these buildings Wimpey decided on extensive use of precasting and set up a 2.3-ha casting yard on the site, with quality control facilities.

During the electrolytic reduction process, the carbon anode is consumed. The smelter complex therefore also includes an anode manufacturing plant for making carbon rods from petroleum coke. There is also a cast house in which the molten aluminium from the pot rooms is cast into various shapes.

In addition to holding contracts for civil, structural and architectural design of the buildings, Wimpey was also responsible for design and construction of mechanical and electrical services around the plant, for site roads, a foul sewerage system, treatment works and outfall plant. A desalination plant was also needed to purify the large amount of water required in the cast house and elsewhere.

The Saudi Arabia–Bahrain causeway

Bahrain has close political ties with Saudi Arabia and in the mid-1970s a plan was developed to build a physical link between the two states in the form of a causeway. This road project, now known as the King Fahd Causeway, was one of the most substantial and ambitious construction contracts in the world at the time.

Outline design was by Saudi-Danish Consultants but the main contract was won, in July 1981, in international competition, by the Dutch firm Ballast Nedam which submitted its own alternative design which used concrete structures and made extensive use of prefabrication. Competitors were astounded by Ballast Nedam's bid which was £33 million below that of its nearest rival and £100 million below the nearest steel bid. Ballast Nedam's contract was valued at £352 million.

The causeway is 25 km long and consists of five prestressed concrete bridges totalling 12.5 km in length, linking seven sections of rock and sandfill causeway and the intermediate island of Umm Nasan. It carries a dual two-lane motorway with hard shoulders.

Construction began in September 1981. Two factors were crucial to Ballast Nedam's strategy. First, it had to assemble a fleet of floating construction plant, much of it specially designed. 'The right large equipment just didn't exist before,' operations manager Pieter Andringa told *New Civil Engineer* in 1982. The fleet included two jack-up barges for drilling the 4-m diameter pile shafts, the 1,500 t floating crane *Ibis*, and the 1,125-kW cutter-suction dredger *Queen of Holland*, in addition to floating batching plants for concrete.

Second, the contractor set up a huge prefabrication yard on an artificial island whose dimensions are 1 km long by 300 m wide, on the northern edge of Umm Nasan. Here all but 50,000 m^3 of the 325,000 m^3 of concrete used in the project was cast. This included the piles on which the causeway is founded, and bridge deck sections.

The piles on which the causeway is founded were the largest ever built. There are 584 of the 3.5-m diameter hollow concrete piles which were cast in sections of up to 6 m, then prestressed together in 20-m to 42-m lengths. Bridge deck sections were cast in two standard lengths: cantilever sections 66 m long weighing 1,500 t, and 34-m long 650-t suspended spans to go between them. A 1,500-t gantry crane on the island was used for lifting.

Purpose-built augers, mounted on jack-up platforms, bored holes up to 4 m diameter in which the piles were to be placed. The holes were protected by steel casings and a 1000-t floating shear leg crane was used to lower each pile into its shaft. Finally the pile bottom and the annulus were grouted.

Piles were placed in pairs at 50-m centres. Only the bridge nearest Bahrain is not founded on piles, because of the risk of penetrating the aquifer which supplies most of Bahrain's fresh water. Instead, this bridge is founded on 20 rectangular precast caissons up to 11.5 m deep and weighing up to 1,000 t the bridge piers are designed to withstand impact from a ship.

Much of the first year of the contract was taken up with temporary works. In addition to the artificial island, a temporary dam was built between Bahrain and Umm Nasan for access. One of the biggest problems was the variable water depth which ranged from 12 m in places to less than 1 m elsewhere. All the floating plant was designed to have a draught of less than 4 m, but even so, considerable dredging was needed to form access channels for the plant.

Embankment sections generally correspond to areas where the water depth is less than 5 m. They are built from two parallel bunds protected by rock armour, infilled with hydraulically placed sand. A total of 5 million t of rock and stones and 10 million m^3 of sand was needed, much of it dredged from the seabed.

The problem of chloride attack on concrete, which caused serious deterioration of many structures throughout the Gulf but especially in Bahrain (see p.64) was by now well known and extensive precautions were taken to combat it. A 720-m^3/d desalination plant was installed to provide concrete mixing water; all sand and aggregate was thoroughly washed. Cement was specially imported from the Netherlands. Prestressing wire was stored in air-conditioned sheds. Concrete from the batching plant was regularly checked to make sure chloride content was below 0.1 per cent. Around the splash zone, concrete piers were coated with epoxy.

The contract was completed several weeks early, in December 1985, though because of delays in letting approach road contracts these were not ready until a year later.

Isa Town

Isa Town was a new settlement similar to new towns built in Britain since World War II. In the Arab world, it was the most ambitious such development in relation to country size.

Its primary aim was to provide housing for low-income groups in Bahrain, since such accommodation was scarce. Provision was also made for medium- and high-income households, to encourage an integrated township.

In all, 2,500 houses with all infrastructure and community facilities were built in five years starting in 1963, for an initial population of 15,000, expected to grow ultimately to 35,000.

Wimpey was recommended by its associate, the Bahraini merchant Yusef bin Ahmed Yanoo. It undertook the design and management of the £6 million contract at a fixed fee. Local contractors were to be used as far as possible to do the actual building work. The Bahraini government provided the site, and paid for the public and administrative buildings. The houses were offered on low-interest 15-year mortgages; some were also made available for rent.

Isa Town is about 3 km from Manama, and the same distance from the centre of the oil industry to the south of the capital. A quick start to building work was required, but not a crash programme: 500 dwellings per year was the target. Within days of the contract being signed, four Wimpey teams were on site: surveyors; laboratory personnel conducting a soil survey; builders and estimators investigating availability of materials and evaluating the best way of providing a water-supply, sewerage and sewage disposal; and architects finding out what sort of town the government and people needed.

After costings and tidal surveys had been done it was decided to pipe the sewage 1.6 km out to sea rather than build a treatment works on land. Because of the flat topography, extensive pumping was needed. Fissured permeable limestone overlies the whole 6-sq km site under a shallow covering of soil. Thus, before trenches, up to 5 m deep, could be formed for drains, the whole area had to be stabilised by pressure grouting.

Trial drilling revealed a suitable water source within easy reach of the site. A supply of 225 l/d per household was allowed, equivalent to UK standards.

Wimpey concluded that local precasting yards could be organised to produce blocks and allow the township buildings to be constructed mainly of concrete blockwork. This is what happened, although there was a contingency plan to change to 'no-fines' concrete if for any reason labour became scarce.

There were three main house sizes and 0.2-ha plots for wealthier merchants to build individually designed villas. The smallest houses were 45 m² in area with provision for a further room to be added without difficulty. All houses were designed to meet the requirements of purdah, with 2.5-m high walls around the courtyards and screening around the roof area, which is used for sleeping. A break with tradition came in the housing density which is much less than the architects originally envisaged because of the need to provide streets for cars. Each house has a parking space; there are secure playing areas for children away from traffic, and traffic-calming measures were used to slow motor vehicles in the streets around the houses.

The town's focal point is the shopping centre with its pedestrian precincts, two large mosques, an open market, and the major public buildings such as the town hall, police station, fire station and sports stadium with an Olympic-standard swimming pool. Shops and workshops are designed on a modular grid allowing them to be adapted to suit small or large enterprises. Other facilities include schools and a library.

Wimpey worked closely with the client during the construction period. Two Bahrain government officials and two Wimpey employees sat on a committee to select subcontractors with suitable expertise. Twelve Wimpey technicians supervised the work of the 2,000-strong Bahraini workforce.

Water supply

Despite the availability of fresh water which migrates to Bahrain in aquifers which stretch under the Gulf from Saudi Arabia, it became clear that as Bahrain developed and its population grew more supplies would be needed.

John Taylor & Sons was commissioned to study the existing distribution system for Manama and Muharraq in 1973. In particular it was asked to study the use of desalinated water from a new multi-stage flash sea water desalination plant, and means of conserving groundwater supplies. It produced a master plan which recommended dividing the area into several distribution zones, each served by a blending station where groundwater and distillate would be mixed. The blended water would be distributed via elevated storage tanks, while underground tanks would store strategic supplies of distillate, groundwater and blended water.

The study included a survey of the existing supply sytem and identified the need for remedial works to reduce losses from the system. A programme of further leak detection studies was recommended, and the consultant's staff trained local technicians to do the work.

John Taylor & Sons followed this two years later with a similar report for the rural areas of Bahrain. It was then commissioned to produce detailed designs and supervise implementation of the plan, which would provide Bahrain's entire population with potable water. Initially this meant providing 120,000 m³/d of blended water for a population of 360,000. After a further strategic report in 1986 the scheme was extended to provide 390,000 m³/d for a population of 700,000 by the year 2000. Around 70 per cent of the blended water was to be desalinated, by distillation and reverse osmosis.

Since the mid-1970s, in a rolling programme of contracts, 15 pumping stations, 17 wellfields, 55 steel ground storage tanks of up to 23 million litres capacity each and 20 welded steel elevated tanks each with 4.5 million litres capacity have been built, together with switchgear, standby power supplies, water treatment works, telemetry systems, and domestic meters. The project's total cost to date is £140 million.

Sewerage

In common with many other Middle Eastern countries, Bahrain realised in the early 1970s that it had an increasing problem of sewage disposal. Through the World Health Organisation (WHO), J. D. & D. M. Watson was appointed in 1974 to draw up a master plan. The study included the old towns of Manama and Muharraq with their more recent extensions, and the new dormitory development at Isa Town. The total population covered was 186,000, expected to treble by the end of the study period in 2008.

An article in *New Civil Engineer* at the time described Bahrain as a 'sewer engineer's nightmare'.[2] The two main centres of population, Manama and Muharraq, are built on a coastal plain and rise barely five metres above the waters of the Gulf. To provide even the minimum sewer gradient would mean deep trenches and frequent pumping stations. But because of the high groundwater table, rich in chlorides and sulphates, deep trenches had been avoided and were virtually unknown in the state. Narrow congested streets meant that access was difficult and there was a danger of undermining adjacent two- and three-storey buildings. Well-point dewatering had recently been used for the first time on Hyundai's ASRY contract.

Existing drainage was primitive, consisting of septic tanks, from which effluent overflowed into concrete-covered earth channels designed to take surface water, discharging to short sea outfalls. Overburdening of the drains was leading to surface flooding by polluted water, and contamination of the foreshore, causing smells and a serious health hazard. Only Isa Town had a modern sewerage system, and even its sewage was discharged to the sea virtually untreated.

Watson decided that the climate ruled against a combined surface and foul sewerage system and proposed a separate sanitary system. Alternatives, such as disposal to sea through a long sea outfall, or treatment of sewage to a standard suitable for reuse, were considered. Surprisingly, however, although Bahrain is an island, there were few possible sites for a sea outfall. Prevailing north-west winds ruled out sites on the north shore; on the east most of the creeks were shallow and congested by causeways and jetties. The south and west coasts were too distant from populated areas to be suitable.

As it happened, however, the Ministry of Works, Power & Water saw sea disposal as waste of a scarce resource. Bahrain's rainfall is minimal, and the aquifers which supplied all its water supplies at the time were threatened with saline contamination because of over-use.

The master plan, therefore proposed three-stage treatment of the sewage, with primary sedimentation, followed by biological filtration and microstraining. The effluent would then be suitable for use in agriculture and on amenity areas.

With the expected increase in population in the major cities, land was at a premium and finding a site for the sewage treatment works was another major problem. Eventually a reclaimed site on tidal flats adjacent to Khor Maqta Tubli, 5 km from the capital, was chosen, even though its size was restricted and prevented treatment processes such as oxidation lagoons from being used.

A characteristic of development in Bahrain has been the amount of land reclaimed from the sea for new projects, particularly around Manama. Many of the new hotels and big buildings, as well as roads, have been built on reclaimed ground. Fortuitously for sewer construction, extensive reclamation was already under way along Bahrain's foreshores, ready for future developments; this provided an ideal opportunity to lay many of the interceptor sewers without causing disruption.

The recommended interceptor and trunk sewerage system comprised 46 km of sewers up to 1.6-m diameter with 34 pumping stations. Because of the risk of attack from high sulphate and chloride concentrations in the groundwater, clayware pipes were used for gravity sewers up to 45-mm diameter and glassfibre-reinforced plastic (GRP) pipes were used for larger sizes and pressure mains. The trunk system became operational in 1978 and was completed in 1979 at a cost of US$85 million.

Watson was given the further commission in 1978 of preparing a master plan for the phased connection of houses to the main sewers, which was to involve providing 110 km of pipe at a cost of US$420 million. By 1985, 113,000 dwellings had been connected to the system.

During the implementation of the master plan, the need to provide groundwater drainage became apparent. There was still significant overflow from septic tanks causing a public health hazard. Initially groundwater flows were taken into the new sewerage system, but to reduce flow to the treatment works to an acceptable level of volume and salinity, a separate ground- and surface-water drainage system was needed.

Contributing to the problem were ageing, defective or overloaded drains, as well as contractors pumping groundwater from excavations. Land reclamation round the coast had cut off existing drainage channels.

A novel solution was adopted to overcome the problem of the low-lying ground. A system was developed in which some sections of the sewer were surcharged, enabling the flow to be discharged to sea without pumping.

In 1984 Watson Hawksley, as it was by then known, received a further commission to review and update the master plan in detail for the next 10 years and in general for the next 25. Investigations included population surveys, measurements of flow (including daily and seasonal variations), tests on sulphide build-up and trials of oxygen and chemical injection to combat this, and closed-circuit TV surveys to establish levels of siltation as well as structural condition.

The strategy eventually proposed envisaged that all major development areas should be provided with main sewerage and treatment facilities, with a substantial section of the total population connected by early next century.

Power supply

Ewbank & Partners and Rendel Palmer & Tritton were active in power supply schemes. Ewbank was responsible for design and supervision of construction of both Manama 'B' and 'C' power stations in the late 1960s and early 1970s. Manama 'B' comprised six 6,100-kW gas turbine generating sets with associated switchgear. Manama 'C' added a further three 13,000 kW generating sets, switchgear and a new control room.

Rendel Palmer & Tritton designed and supervised the 100-MW gas turbine power and sea water desalination plant at Sitra which was built on reclaimed land 1 km offshore. The project included construction of a 1-km marine causeway and 6-ha site, as well as dredging of a supply channel for cooling water.

Rendel Palmer & Tritton also took part in a study of long-term power and water requirements for the state and made proposals for the phased construction of the Sitra station. Under the improved power distribution system for Bahrain, the firm was also responsible for three 66-kV substations.

Ports

At Mina Sulman, Rendel Palmer & Tritton was consultant for a development which more than doubled the area of the existing port and increased cargo handling capacity threefold. From 1974 onward, it provided initial technical and economic studies and prepared the development master plan, followed by design of civil, structural, electrical and mechanical works, and supervision of construction.

In the first development stage, completed in 1979 at a cost of £62 million, six new deepwater berths each 200 m long by 300 m wide were provided, with cargo handling facilities and quayside services. A small craft harbour, new port buildings, and considerably improved access from both road and sea were included. More than 40 ha of land was reclaimed, and there was provision for further reclamation and construction of another 10 or more berths.

In stage two, completed in 1984, the port was modified to provide container handling and stacking areas and transshipment facilities. Two 40-t cranes were provided. Construction value was £14 million.

Concrete deterioration in the Gulf

An unpleasant surprise for engineers working in the Middle East was the discovery of the exceptional speed at which concrete could deteriorate under certain conditions. Nowhere was this problem more severe than in Bahrain.

Concrete began to be used in Bahrain during the 1950s, earlier than in many other Middle Eastern countries, and by the 1970s deterioration was already beginning to show in some structures.

A number of causes combine to attack concrete in the Gulf. Reactive aggregates, the hot climate, and water are all implicated. There is no single mechanism for deterioration, but rather several which interact and feed on one another. However, the key to the problem is chlorides and sulphates which can be present in groundwater and aggregates. Chlorides are also found in the atmosphere near the sea.

Conditions in Bahrain are particularly severe because its water-table is close to the surface. Chlorides – found in common salt and hence in sea water and often in groundwater, particularly in coastal areas – set up an electrochemical reaction with the steel reinforcement, rapidly causing rust. Sulphates attack the cement.

Alkali-aggregate reactions occur between reactive coarse aggregates, such as dolomite, and alkalis present in cement or in sea water which penetrates the concrete. Alkali-silica reaction forms silica gel on the outside of the aggregate – this causes a large increase in volume and forces areas of the concrete surface to break away.[3]

How do these unwanted chemicals get into the concrete? Firstly, they can be introduced into the mix itself either through the use of saline water or through the aggregates. Where the water-table is close to ground level, moisture can rise through sand to the surface by capillary attraction, where it evaporates, depositing salts behind it. Although the sand looks normal it is contaminated with these deposits. Chlorides can be removed by washing, but sulphates cannot, so the only solution is to discard the top layer of sand and take it from below.

Similar contamination can occur on coastal plains which are periodically flooded and where much of the development in the Gulf is sited. Limestone, a common aggregate throughout the Middle East, can have its upper layers contaminated in a similar way.

This means that great attention to detail and extensive aggregate testing are needed during quarrying to ensure that contaminants are not introduced. Similarly, reinforcing steel must be thoroughly cleaned before use.

Even if extreme care is taken to ensure that no potential contaminants are contained in the ingredients of the concrete, salts carried by groundwater, or in the humid atmosphere, or from sea water flooding, can find their way in, particularly if there are cracks. In the Gulf, if newly placed concrete is not protected, high temperatures and winds can cause it to dry out at the surface, leaving shrinkage cracks. Continuous thermal expansion and contraction can have the same effect. There may be a difference of 13°C between day and night temperatures. And in the high temperatures of the Gulf, the rate of any incipient reaction is accelerated considerably.

These effects can be combated by taking extreme care during concreting, shading both the batching plant and placed concrete, and cooling the mix water to keep down the temperature of the concrete in order to limit shrinkage. Concrete foundations in salty ground must be protected with bitumen tanking to prevent attack.

Among the casualties of this process in Bahrain were the Gulf Hotel and Government House. Defects were first noticed in the hotel extension in 1974 and were put down to chloride contamination of the aggregates. The original block used aggregates from a different source and escaped unscathed. However the hotel was closed for 18 months while Wimpey carried out repairs using epoxy resin injection, monitored by consultant

G. Allen Herbert & Partners. At the time the hotel's owner absolved the contractor from negligence, but Wimpey did not charge for the £3.5 million worth of repairs and paid £1 million compensation.

Another structure to suffer was the causeway linking Manama and Sitra, designed by Rendel Palmer & Tritton. This was a 4-km rockfill structure with two bridged spans, the 18-span north bridge being 216 m and the 48-span south bridge being 576 m long. The causeway carries a four-lane dual carriageway, plus power cables and main water pipelines across the tidal channel between Sitra and Bahrain island. The bridges were incorporated in order to prevent scour and to minimise disturbance to shrimp beds exploited by the local fishing industry.

The bridges are of composite construction, with a reinforced concrete deck on steel beams. Foundations consist of tubular steel piles with concrete pile caps.

In 1986 an inspection revealed cracking in the road surface of the bridges. Further investigation, including trial excavation and taking of core samples, revealed severe concrete spalling and corrosion of the upper reinforcement. Preliminary tests showed high chloride levels.

The client, Bahrain's Ministry of Works, Power & Water, re-engaged Rendell Palmer & Tritton to assist in investigating the problem and to recommend ways of restoring the structure to its full load-capacity.

Detailed inspection and tests confirmed that cracking and lamination in the deck, and corrosion of the reinforcement, were widespread. The cause was identified as chloride ingress – the source was thought to be sea water spray, and chloride-rich dust lying on the surface and which was being drawn into the concrete after dissolving in dew or occasional rain.

At the time of the causeway's construction in the mid-1970s, the problems of concrete deterioration due to salts in the materials were well-known and the concrete specification took this into account. In fact, says Graeme Marshall, head of the bridge department for Rendel Palmer & Tritton: 'The specification was ahead of its time in terms of preventing that problem, and it succeeded. But the problem which has occurred, that of chlorides getting in from the outside, wasn't understood at the time.'

A precautionary load restriction was imposed on the causeway during further studies. Rendel Palmer & Tritton assessed the structural adequacy of the bridges and concluded that the deck remained adequate for its design axle load of 22 t but was overstressed by 8 per cent by the design local wheel load of 30 t.

Various options for the deck's repair or replacement were considered. Repair possibilities considered were: a minimum or patch repair option; repairing the entire deck surfacing to below reinforcement level, with the addition of deck waterproofing; and the second option plus cathodic protection. This final option was thought to be the most satisfactory for providing the greatest durability with the minimum of future maintenance.[3]

Building a new deck on the existing steel beams, which were in good condition, was considered the best replacement option. The new deck would be thicker, giving greater top cover to the steel, with higher quality concrete, a waterproof membrane, a revised cross-section to improve drainage, and improved joints between the ends of deck slabs.

A cost comparison, assuming that a repaired deck would last 15 years and a new deck 30 years and taking into account an extra 14 weeks' traffic disruption which total replacement was estimated to entail, found that the cost of each option was similar. Rendel Palmer & Tritton recommended replacement.

However, the ministry also asked Rendel Palmer & Tritton to suggest temporary strengthening measures for the deck so that replacement could be deferred for as long as possible. With the top 40 mm of concrete ineffective, which was considered to be the worst condition to which the slab could deteriorate, the deck could be shown to be structurally adequate, though it failed to meet serviceability requirements. Rendel, Palmer & Tritton therefore proposed a system of propping, or K-bracing, to the slab only in places where there were also defects on the underside. Some patch repairs were carried out, and a system of three- and six-monthly inspections, plus detailed monitoring of cracks in selected areas, was set up to monitor further deterioration and extend the propping where necessary.

The reinforced concrete pile cap beams were also at risk from chloride attack. Because the reinforcement here had greater concrete cover, however, serious deterioration had not yet begun. Here Rendel Palmer & Tritton decided to add a system of cathodic protection – which according to research by the US Federal Highway Authority had concluded 'the only rehabilitation technique that has proven to stop corrosion in salt-contaminated bridge decks, regardless of the chloride content of the concrete'. The chosen type of cathodic protection employed a titanium mesh with a polymer-modified mortar-sprayed overlay.

Following the temporary repairs and instigation of the monitoring programme, 'the deck didn't deteriorate to the extent we expected,' stated Marshall. The cathodic protection to the pile caps also proved successful. 'Cathodic protection technology has also moved forward a great deal in the last few years,' says Marshall, and as a result Rendel Palmer & Tritton 'looked at the repair plus cathodic protection option in a lot more detail'. The outcome was that its recommendation to replace the deck was changed.

Instead, in April 1993, the contractor Morrison Construction began work on a 12-month, £2.25 million contract to repair the deck. Morrison was to remove the asphalt wearing course and spalled concrete by hydrodemolition to a depth of 40 mm over the entire surface area. The slab and the wearing course were both to be replaced in concrete, giving more cover to the wearing course. And cathodic protection was to be provided by installing a titanium mesh in the top 30 mm of the slab. This was expected to extend the life of the structure for another 15 to 20 years.

This is not the only example of concrete deterioration in the Middle East. It is clear, however, that the experience gained has greatly increased understanding of concrete durability, the benefit of which will be felt worldwide.

Bahrain today

The first Gulf state to exploit oil will also be the first to exhaust its resources. But Bahrain has planned for this eventuality; assisted by the fact that it has a relatively highly educated population, it has had more success than most of its neighbours in diversifying its economy. With a sophisticated industrial base, and its success in attracting financial service industries, Bahrain can be confident of prosperity continuing after oil runs out. Politically, Bahrain has a history of unrest; the ruling family may find itself under pressure to introduce more democratic structures in the future. Continuing opportunities for western construction firms are likely to be modest.

chapter five

SAUDI ARABIA

Nowhere did oil wealth create transformation on so vast a scale as in Saudi Arabia. In the 1940s the impoverished country was discovered to be endowed with a quarter of the world's oil resources, just as demand in Europe and the US was set to reach new heights as reconstruction after World War II began. Development in the neighbouring United Arab Emirates was spectacular, but the emirates are tiny by comparison: Dubai, for example, has an area of only 3,900 sq km compared with Saudi Arabia's 2.24 million sq km.

With the quadrupling of oil prices in 1973, even the enlargement of Saudi Arabia's already ambitious programme of infrastructure works could not keep pace with the influx of revenue. In 1975 the Kingdom was receiving as much every hour as it had in a year in the 1930s. In the decade to 1985 Saudi Arabia not only completed basic infrastructure such as the road network, it also built 25 airports, two huge industrial cities (Jubail and Yanbu) created on greenfield sites at a cost of US$70,000 million, and even encouraged agriculture to the extent that it became self-sufficient in wheat.

The Kingdom of Saudi Arabia can trace its origins back to 1745 when the chieftain Muhammad ibn Saud formed an alliance with Muhammad ibn Abd al-Wahhab. The latter was the founder of the Wahhabi movement, an austere and strict version of Islam intended to be a return to the religion in its original state, stripped of inessentials added since the time of the Prophet. Propagation of this idea was the original purpose of the creation of the Saudi state. It remains central to Saudi Arabian culture and descendants of Muhammad ibn Abd al-Wahhab continue to occupy influential positions in the Kingdom today.

Most of what is now Saudi Arabia was not part of the Ottoman empire. Only the Hejaz, the western coastal plan which includes Mecca, Medina and Jeddah, came under Turkish rule.

Wahhabi rule spread in the eighteenth and early nineteenth centuries. In the early 1800s the Wahhabis seized the Holy Cities of Mecca and Medina from the Ottomans. This incurred Istanbul's displeasure, and after a seven-year campaign, the Wahhabis were defeated by a

Turko-Egyptian force. The Amir, Abdullah, the son of Ibn Saud, was executed.

Wahhabi power rose again between 1824 and 1865 under the rule of a grandson of Ibn Saud, Turki, and of Turki's son Faisal. But following Faisal's death there ensued 25 years of civil war between two of his sons, culminating in the capture of Riyadh, the Saudi capital, by the rival Rashid family. In 1891 a third son, Abdul Rahman, went into exile in Kuwait, and it was from there that the modern Saudi Arabian state was launched at the beginning of the twentieth century.

Abdul Rahman's son, Abdul Aziz, who later took the name Ibn Saud, was responsible. He was an inspirational leader, but one who was aware of his weaknesses as well as his strengths. He was able to deal diplomatically with world leaders and oil companies when it became necessary.

In 1902 Ibn Saud seized Riyadh in a surprise attack. He was to rule the Kingdom he set about re-creating until his death in 1953. Ibn Saud fought his main rivals, the Rashidis, for the next 20 years before finally subduing them and capturing Hail, to the north, in 1922.

At the same time as Ibn Saud was setting about restoring his kingdom, Istanbul was trying to consolidate its hold on the area. In 1900 the Ottoman Sultan, Abdul Hamid, concluded an agreement with Germany to build a railway line through Palestine and the Hejaz, from Damascus to Mecca, ostensibly to facilitate the transport of pilgrims. The only railway lines in the Ottoman empire east of Istanbul at the time were from Damascus to Beirut and from Jerusalem to Jaffa.

In reality, the railway formed part of a Turco-German grand design to extend their influence into Arabia and Mesopotamia (Iraq), where a parallel project was begun to build a railway from Aleppo (Halab) through Mosul and Baghdad to Basra.

By 1904 the first 460 km had been built from Damascus to Maan, near Petra ('the rose-red city half as old as time') in Jordan, and by 1908 it had reached Medina, 1,320 km from Damascus and only 450 km short of Mecca.

When T. E. Lawrence (popularly known as Lawrence of Arabia) joined with the Hashemite Arabs of the Hejaz to fight the Turks, one of their principal targets was the Hejaz railway, and their exploits are vividly recounted in Lawrence's *Seven Pillars of Wisdom*. So successful were their raids that the Saudi Arabian section of the line was not rehabilitated after the war and has lain unused ever since – although there was an abortive British initiative to reconstruct it two decades ago.

The Hejaz became part of Saudi Arabia in the 1920s, after conflict broke out between Ibn Saud and Sharif Hussein of Mecca. This culminated in the fall of Jeddah and Medina in 1925–6, giving Ibn Saud control of the Hejaz on the west coast. In 1929–30 Ibn Saud defeated a rebellion by a warrior brotherhood known as the Ikhwan and consolidated his territories into the Kingdom of Saudi Arabia. A war with Yemen in 1934 confirmed his hold on territory to the south-west and ended his acquisitions.

At about that time the discovery of oil in Bahrain sparked interest in exploration along the whole of the east coast of the Arabian peninsula. Facing a severe shortage of funds, Ibn Saud entered into negotiations with Standard Oil of California. Seeking to exclude competition, the Anglo-Persian Oil company also made a bid but was prepared to offer a much smaller advance payment. A concession covering the whole eastern part of the Kingdom was granted to Standard Oil in 1933, marking the beginning of close links between Saudi Arabia and the US.

Oil was discovered in commercial quantities in Dammam in 1938. But this was just the start. Further exploration in the 1940s led to the discovery of the Abqaiq field, bigger than any in the US. This coincided exactly with the emergence of the US as a net importer, rather than an exporter, of oil. Though World War II was still in progress and materials were in limited supply, scarce steel was made available to construct a refinery. In 1945 it was possible for oil production to begin on a serious scale. As Kuwaiti oil also came on stream at about the same time, the proportion of the world's oil produced in the Middle East rose to 12 per cent, increasing to 25 per cent a decade later.

Meanwhile, in 1944, Standard Oil, which had been working in joint venture with Texaco, sold shares in its concession to Exxon and Mobil to form the Arabian-American Oil Company (Aramco). Aramco supervised two major projects in the 1950s, the Trans-Arabian Pipeline (Tapline) linking the newly discovered oilfields in Saudi Arabia to the Mediterranean, and the Riyadh–Dammam railway.

Ibn Saud died in 1953 and was succeeded by Saud, his eldest son. Initially the Saudi Kingdom's oil wealth brought little benefit, since both Saud and his father had little concept of planned spending. Saud modernised the Kingdom's government, introducing a council of ministers, but his administration was so disorganised that by the late 1950s it was facing financial collapse and had to seek an International Monetary Fund loan. Prince Faisal was made finance minister and temporarily restored order until Saud took control again himself. The Kingdom hovered on the edge of crisis until Faisal was made prime minister in 1962. Two years later, Saud was persuaded to hand over power completely.

Having travelled in Europe and the US Faisal was more worldly-wise than Saud. He had a vision of how the country should develop which has remained the Kingdom's guiding principle ever since: he wanted to modernise it physically while retaining its traditional social and religious values. In this he seems to have succeeded, and under his rule he brought stability to the Kingdom.

As oil income grew, the Kingdom embarked on a programme of infrastructure works and the building of schools and hospitals. Among the innovations that Faisal introduced were television, and education for girls, both of which brought storms of protest.

Under Faisal's sensible economic management Saudi Arabia was, by the early 1970s, building up a large financial surplus. There was heavy spending on infrastructure, education and health, while around a quarter of the Kingdom's revenue went to defence. Much of the population was still living frugally, but development could not keep pace with increasing revenue.

The 1973 Arab–Israeli war transformed the situation totally. Arab oil producers announced a boycott of sales to countries supporting Israel and cut back production. This led to a quadrupling of the price of oil to US$15 a barrel. When production was restored to normal levels in 1974, revenue rocketed. It increased even further over the next two years with the gradual nationalisation of Aramco and further large increases in production. Revenue rose from US$2,700 million in 1972 to US$25,000 million in 1975.

There were four immediate consequences of this new-found wealth. First, the prospect of completing the country's infrastructure by the end of the decade became a reality. The Kingdom's second five-year plan, from 1975–80, envisaged spending US$142,000 million – ten times the amount allocated in the first plan.

A second fundamental objective was to use oil as the basis for developing other industries. Two huge and entirely new industrial cities were to be created at Jubail and Yanbu at an estimated cost of US$70,000 million as the linchpin of this strategy. Third, immediate benefits were conveyed to the Saudi Arabian people through subsidies and the removal of taxes.

Fourth, Saudi Arabia became a major international aid donor through participation in multilateral agencies and the activities of its own Saudi Fund for Development (SFD), set up in 1974. During the

1970s the Kingdom spent over 10 per cent of its gross national product on aid.

What were conditions in the Kingdom like for construction firms from overseas? In 1975, Hugh Try of the contractor W. S. Try went on a fact-finding mission to Middle East states, investigating prospects for winning overseas work for the first time. The tour took in Saudi Arabia, Dubai, Sharjah, Abu Dhabi, Oman and Qatar. He described his conclusions in a paper to an ICE conference in 1984: 'Saudi Arabia was the biggest construction market, least well served by international contractors or local contractors. It was, however, clearly the most difficult market in which to operate, with the most difficult physical conditions, the least developed infrastructure, and with poor backup in terms of hotels, telephones, port facilities, etc.' However, in the other markets competition was already well established. 'We therefore decided that Saudi Arabia represented the best opportunity . . . in the sense of being the place where we would be least disadvantaged by being a newcomer.'[1]

In 1975 King Faisal was tragically assassinated in a revenge attack by a deranged nephew. He was succeeded by his brother Khaled, though the main burden of administrative duties fell to Crown Prince Fahd, who became first deputy prime minister.

Khaled's reign, which lasted until 1982, was a golden age for Saudi Arabia. Under the second five-year plan, the Kingdom built ports, airports, power stations, roads, desalination plants, schools and hospitals. It set up the national airline Saudi Arabian Airlines (Saudia), and built the basic infrastructure for Jubail and Yanbu, supervised by a Royal Commission.

Aramco oversaw the construction of the US$1,000 million Master Gas System and an expansion in oil production capacity to 12 million b/d. Interest-free loans were made available for housing and property, and low interest loans were provided for setting up light industry in the private sector – evoking an enthusiastic response from businessmen, most of whom set up plants to manufacture building materials.

The third five-year plan ran from 1980 to 1985. It essentially finished development of the Kingdom's infrastructure, and provided for a big expansion in the number of state hospitals. A key element was the setting up of petrochemicals plants and oil refineries at Jubail and Yanbu. A surprising achievement was in agriculture where, irrigation schemes arising from the study of the Umm er Radhuma aquifer and a government policy of buying indigenous wheat at six times the world price, led to the Kingdom producing a surplus. This continued even after the price was halved. Saudi Arabia also became self-sufficient in poultry and milk.

But during this period the golden age came to an end. In 1981 new rules came into force which required all consultants and contractors to bid formally for work; and any bidder had to have a Saudi Arabian partner. Previously work had often been given to the firm with the best track record with little in the way of contractual documentation. The new system did 'more harm than good to both sides,' said a western engineer quoted in *New Civil Engineer* in 1982. Unscrupulous consultants 'from countries without a watchdog institution looking over their shoulders' submitted conceptual drawings rather than complete designs, he claimed. The contractor then ended up 'doing most of the design because the specification was imprecisely worded'. Consequently an inexperienced contractor took 'twice as long'.[2]

Other firms said that they were no longer able to offer the 'enormous breadth of service' under which much peripheral work would effectively be carried out free. 'Now we do no more than our precise instructions, otherwise we're not commercial,' said one UK firm at the time. 'Goodwill has gone out of the window.'

Halcrow, one of the best established UK consultants, was affected as much as any other firm. It had been working on harbours in Jubail, Yanbu, Jeddah and elsewhere for 17 years, winning work often without competition, from the Saudi Ports Authority on the strength of its reputation. Suddenly the flow of work dried up. It had to cut back on staff in an attempt to remain competitive.

In 1985 Saudi Arabia's oil production dropped to just 3 million b/d partly because of a drop in world demand and partly through Saudi Arabia's role as 'ruling producer', in which it limited its output to keep the total production of OPEC countries within the ceiling OPEC set. Oil prices fell sharply in 1986, and the fourth five-year plan, which was intended to encourage the private sector to plan a greater role in industrial development, was almost stillborn. The construction industry was badly affected, with many local firms going bankrupt and overseas firms largely pulling out.

The recession nevertheless had the benefit of reducing expectations of high income and brought about a reduction in industrial operating costs. The economy remains driven by government spending, with most private sector firms existing to serve the demands of the government rather than to trade between themselves.

Within the last few years there has been something of a revival of interest among construction firms in working in Saudi Arabia, though many find that intense competition and fee bidding make it impossible to win straightforward work at an economic rate.

Some have expressed doubts about the continued health of the Saudi Arabian economy, however. Partly because of the cost of the Gulf War, reserves have been cut to a tenth of their level a few years ago. The Kingdom's expenditure on arms and defence works remains high. Under the Al Yamamah agreement between the Saudi and British governments, British Aerospace was to supply £5,500 million worth of military aircraft and support facilities. In addition to aircraft and mine-sweepers it included plans for an airbase worth £10,000 million. The term 'airbase' is inadequate to describe the complex which would have covered 900 sq km, comprising a 25-sq km airfield with three 4-km runways and a town to accommodate 25,000 people. Consultants W. S. Atkins, Sir Alexander Gibb & Partners and Ewbank Preece & Partners had worked with Ballast Nedam (then a British Aerospace subsidiary) on Master Planning and Sir Alexander Gibb & Partners was working on detailed designs, with British contractors expected to win most of the construction work, when the base was shelved indefinitely in 1992.

Ports

In the early 1970s, a gigantic infrastructure building programme formed the central thrust of Saudi Arabia's second five-year plan. It rapidly became obvious that expanding port capacity was crucial, because without this it would be physically impossible to import the vast quantities of building materials and plant needed for the other projects. Congestion at the ports was already serious, with long queues of ships waiting weeks to unload their cargoes.

In 1976 Dr Fayez Badr became president of Saudi Arabia's port authority. He quickly gained a reputation for business-like efficiency as well as for being quite a hard client master. Badr set up a crash pro-gramme to tackle the congestion problem. He began by banning vessels over 15 years old unless they had certificates to prove that their unload-ing equipment met certain standards, and he reformed the priority berthing system to give preference to ships with the most efficient unloading gear. The authority also encouraged pre-slinging, bagging or palletisation of cargoes to expedite unloading.

Badr then realised that this might simply move the bottleneck inland to the ports' warehousing. Merchants, short of warehouses elsewhere, were using transit sheds at ports to store goods for months or even years. Badr insisted that any goods left in the port for more than a fortnight would be taken to a special compound where they would be auctioned if not collected. This strategy met with considerable success.

At the same time the ports themselves were expanded at an ever-increasing rate. Jeddah, on the eastern shore of the Red Sea on the trading route between Suez and Aden, was the principal port of Saudi Arabia and its main trading centre. The first modern port facilities were set up in 1948 by the far-sighted King Abdul Aziz. In 1964, because of concern about delays caused by large numbers of pilgrims bound for Mecca passing through the port, Sir William Halcrow & Partners was appointed to report on its future development.

In 1966 work was put in hand to implement Halcrow's recommendations, to construct eight deepwater berths dredged to 11 m and capable of handling 1.5 million t of cargo a year. The following year the scope was expanded to include covered barge and lighter wharves and two additional deepwater berths, increasing the port's total area to 135 ha of which about half was on reclaimed land. The expanded port was inaugurated on 1 January 1973 by King Faisal. But even then it was clear that further expansion was needed, leading to phases II and III of the expansion programme. These were put in hand simultaneously for completion in 1979.

The new expansion provided seven new general cargo berths dredged to 12 m, roll-on roll-off (ro-ro) facilities dredged to 8 m, with a total quay length of 4,500 m and a further 368 ha of reclaimed land. Meanwhile cargo throughput rose from 1.2 million t in 1973 to 2.8 million t in 1975 and was continuing to increase at a rate of 50 per cent a year. Phase IV, begun in 1976, added yet another 20 berths: two ro-ro, two for container traffic and the rest for general cargo.

A year after the development of Jeddah began, in 1967, Sir Bruce White, Wolfe Barry & Partners was appointed to extend the port of Dammam on the east coast.

Sir Bruce White Wolfe Barry's resident partner in Saudi Arabia, Roy Cooke, told *New Civil Engineer* some years later: 'We thought it a good job to land, but one of a fairly average size.'[3] By 1977 Dammam port had expanded into a US$3,000 million project with 33 deepwater berths which was to run for many years to come. As a harbour development it was one of the biggest ever. The contract for the last 16 berths, awarded to West Germany's Philip Holzmann, Archirodon of Greece and Interbeton of the Netherlands in 1976 and worth £660 million, was said at the time to be the biggest single contract ever placed.

The port had modest beginnings. In 1967 it was a single jetty 8 km offshore reached by a railway trestle. (It is located out to sea because of Saudi Arabia's gently sloping eastern shoreline: it is necessary to go far out to reach deep water.) By the time development was complete the port had grown into a reclaimed island, 4 km by 5 km.

Initially, however, Sir Bruce White Wolfe Barry was asked simply to extend the trestle to form a dedicated fertiliser berth. By the early 1970s it was apparent that something more was needed. Congestion at Jeddah, the traditional port for importing to Riyadh, was so severe that it was decided to develop Dammam as an alternative. Sir Bruce White Wolfe Barry was asked to undertake a study and prepare a master plan. 'The study produced virtually what we see there today,' says David Trumper of Acer Sir Bruce White (formed from a later merger).

One of the first tasks was to build a dual carriageway which would eventually lead all the way to Riyadh. A causeway was formed to one side of the existing railway trestle by tipping rock to form a break-water and reclaiming land behind it.

The whole area of the port was dredged from an original depth of between 3 m and 7 m to the required 15 m. The dredged material, mainly sand of various grades with lenses of metamorphic rock, was used to build the island forming the port. A 5-km channel was dredged out to deep water. Depth at berths varied depending on the type of craft they were to serve.

The east side of the port was the first to be developed. In theory this took place in four phases: phase I was berths 1 to 12; phase II, the small craft harbour and approach causeway; phase III the administrative buildings; and phase IV berths 13 to 22. In fact, says David Trumper, these phases became merely an administrative convenience: 'In practice, they were all built at once.'

The initial study took two years. Following some delay by the government in awarding the contract, most of the construction of the east port was undertaken in 1971–5 with Archirodon as the principal contractor. The port began operating, though incomplete, in 1976.

Reflecting the experience of Sir Bruce White himself with the design and construction of the Mulberry Harbours used for the D-Day landings in Normandy, the quays were built of 20-m long and 17-m high concrete caissons, prefabricated on site, floated out and sunk into position. They were manoeuvred by three tugs, two pushing on faces at right angles and one pulling diagonally. There was radio communication between the tugs, the shore and the caisson; once the caisson was in the correct place it was sunk by a system of pumps on the caisson itself.

The next major development, which began at around the same time, was the construction of the western port, which continued until 1982. A joint venture of Holzmann, Archirodon and Interbeton won the contract for the basic infrastructure: the wall face, bunkering facilities for fuel and water, a railway system, surfacing, sewers and transit sheds. As the port progressed, smaller 'infill' contracts were let, for example for water supply, or an explosives berth.

A huge workshop was included to supplement the original one which had been designed to serve only the first group of berths. There are in fact two ship repair yards – one for ships using the port and one for the port's own vessels (small craft such as tugs), also designed by Sir Bruce White Wolfe Barry. This port's yard has a 1,500-t shiplift which places the boat on a system of rails so that, once on dry land, it can be moved around the yard.

The main infrastructure of the western port was completed by 1982, though additions continued to be made. There is a container terminal, as well as the usual administrative buildings such as customs offices, transit sheds and a 5,000-t, completely automated cold store. Trains, 3-km long, leave every day laden with port cargo.

Dammam was one of the biggest port developments ever, but it was surpassed by that of Jubail.

Jubail - a new industrial city

In the early 1970s, plans were initiated for a massive new industrial city and port at the site of a small fishing port, Jubail, on the Gulf coast. It was one of two developments, the other being at Yanbu on the Red Sea, which were intended to provide Saudi Arabia with an industrial base which would help to diversify exports and guarantee income when oil reserves were exhausted.

A Royal Commission was set up to oversee the project, and Bechtel was appointed as project manager with a construction management team which peaked at over 600 strong, responsible to the Royal Commission for the execution of all aspects of the master plan, for detailed project planning, and for supervision of an international group of consultants responsible for the detailed design. Bechtel was not allowed to bid for detailed design itself. Its London office coordinated the project partly because it was the Royal Commission's wish that most of the infrastructure work should be done by UK consultants. The master plan was breathtaking in scope, envisaging an integrated natural gas complex, an iron and steel works, an aluminium smelter, two oil refineries and three petrochemicals plants.

Roger Dobson, now director-general and secretary of the ICE, was at the time Bechtel's chief civil engineer. Building Jubail, he says, was the equivalent of building the industrial complex at Middlesbrough on Teesside (in north-east England) in 10 years:

> The types and spread of industry, and the population, were similar. But Middlesbrough took 150 years to develop from a north-east coast fishing village up to a major industrial centre, starting in the 1820s and peaking in the mid-1960s to 1970s. San Diego, the next comparable place, took about 50 years. Jubail took 10 to reach the same stage of development in terms of industry, population and wealth creation, from very similar beginnings. Jubail and Middlesbrough were fishing villages, San Diego was a hamlet down on the south-west coast.

Most of the industry was built during the third five-year plan. During the second plan, the basic infrastructure was provided – itself no mean achievement. Priority was given in the first instance to completing a deepwater port, and desalination and power plants simply for the benefit of the project itself. Work on the port alone, costing £1,600 million, included building 20 km of causeway and breakwater, 37 million m^3 of dredging, 11 million m^3 of imported fill, and 3.3 million m^3 of concrete.

A wide range of major British consultants were involved in the project. Sir William Halcrow & Partners was appointed to design and supervise the port itself. Binnie & Partners worked on water supply, sewerage and sewage treatment, while Sir Alexander Gibb & Partners was later responsible for the seawater-cooling system and the airport.

Feelings about working under Bechtel were at times mixed. 'It was a very interesting clash of cultures,' says Roger Dobson. 'UK consultants were used to being commissioned by clients as "the Engineer". They expected to be trusted – the mere fact that they were given an assignment, and their reputation, assured people that it would be delivered.'

In terms of control Bechtel ran the Jubail project as if it were a process plant. Its system included strict monitoring of progress of the design, which some consultants thought excessively bureaucratic or a waste of time.

Drawing controls were a typical example. There would be a list of which drawings were to be submitted, with details of what would be on each drawing, with a start date, an end date and a number of labour hours. Target dates for reaching milestones for each drawing – 30 per cent, 70 per cent, 90 per cent complete – were also set. 'There were certain UK consultants to whom this was complete anathema,' says Roger Dobson. 'They were not used to accounting for their man-hours on a weekly basis, or accounting for drawings or designs in terms of per cent complete.' But, he adds, the flow of information was so crucial to the overall plan that it was essential to do it this way.

Although there were difficulties and some firms were never happy with the arrangement, Roger Dobson says, however: 'It did some of the consultants a power of good. They were the first to admit it afterwards, though nothing would have persuaded them to say anything good at the time. What the system does is make people very disciplined in their production of information, and it highlights problem areas very early in the project.' A similar philosophy was used more recently to pull back delays on London's Limehouse Link tunnel.

The port of Jubail had a twofold function: apart from serving the new industrial city it was to be a general cargo port to relieve congestion at Dammam, 80 km away.

Jubail was well placed for the type of development envisaged, being a convenient centre for the purification of natural gas, which would be the main fuel for the new industries. It also offered a site sheltered from the north by the Ras Abu Ali promontory. A disadvantage, as with Dammam, was the gently sloping western Gulf shore, not particularly suited to the construction of a port. However, at a distance of 10 km offshore the water depth increased to 30 m.

The port has three parts: the commercial harbour, the industrial harbour, and an open sea tanker terminal, outside the harbour basin but connected to the mainland by a causeway, with four berths in water depths of over 28 m.

The two harbours have separate 14-m deep entrance channels. The common basin encloses 4,000 ha of water. The main commercial quay is 1,760 m long overall and 600 m wide, with eight berths in a 12-m depth of water. A further 2,100 m of quay provides another nine berths in 14 m of water, with provision for containers.

The breakwater to the north of the harbour also acts as a causeway, 300 m wide and 9 km long, giving access to the industrial harbour and the open sea tanker terminal. It carries a road, pipelines and conveyors. The end of the causeway is so far out to sea that from its end the coastline is no longer visible, while the commercial harbour is a speck on the horizon to the south.

The industrial harbour comprises a 50-ha main quay with nine 14-m draught berths, approached by a dredged channel 14 m deep and 500 m wide. A second quay is provided with a ship repair yard, ship lift facility, and a general maintenance workshop. The industrial complex itself is sited near the shore end of the causeway.

From the end of the causeway, a 3-km steel-piled trestle carrying a road and a two-tier pipe track leads to the four berths of the open sea tanker terminal. These lie in water over 28 m deep and can accept tankers of 300,000 dwt. Each berth has a 40-m by 60-m loading platform. A total of 100,000 t of steel was used in building the trestle. It is used for export only, since oil is supplied to Jubail's refineries by pipeline, and it can handle over 40 million t a year.

Construction got under way in June 1975 with the award of two dredging contracts. This should have been a single contract, but no tenderer could undertake to finish the work in less than five years, whereas the client wanted completion by mid-1978. It was therefore split into two and awarded to the two lowest bidders, both Dutch. Adriaan Volker Dredging took the commercial harbour and first 3 km of the industrial causeway at a price of £165 million, while the Jubail Harbour Consortium, comprising Stevin Dredging, Boskalis Westminster and Zanen Verstoep was awarded the rest of the industrial harbour for £235 million. At the same time the Greek contractor Archirodon was awarded a £75 million contract to begin work on the breakwaters and complete two deepwater berths by 1977, as a step towards relieving congestion elsewhere.

Both Adriaan Volker and the Jubail Harbour Consortium began using cutter-suction dredgers but discovered that the caprock was both deeper and stronger than anticipated. Volker persevered with its dredgers, though maintenance and downtime to replace cutter picks were considerable. In the industrial harbour the rock was even harder and the Jubail Harbour Consortium decided to use explosives. Toa Harbour Works, subcontractor to Archirodon for dredging, used explosives from the start.

As was usual in the Gulf, good quality rock was hard to find. Ordinary rockfill for reclamation came from a quarry 27 km away. Higher-quality rock for armouring the breakwaters was brought from a specially opened quarry in the emirate of Ras al-Khaimah. Aggregate was tested for strength and salt content, and stored on concrete pads to prevent salt contamination.

The main contract for the commercial harbour, worth £520 million, was won by a consortium of Royal Adriaan Volker, Hochtief of West Germany and Consolidated Contractors Company of Lebanon.

The South Korean contractor Dong Ah, subcontractor to Adriaan Volker for imported fill, won praise for the efficiency of its earthmoving operation. *New Civil Engineer* quoted Dong Ah's coordination manager Hyun Ja Ku at the time as saying: 'Most of our 340 men are ex-army. They are used to discipline, so we treat them like the army.'[4] Discipline included 12-month tours of duty without holidays, working 13 days a fortnight. On the morning of the fourteenth day, the drivers paraded for inspection of their trucks.

This contract gave one of the first signs of the military efficiency of South Korean contractors; it was to become a byword as South Korean firms' extremely keen pricing won them more and more contracts over the following years throughout the Gulf.

Apart from the port itself, the biggest piece of infrastructure associated with the industrial complex was the seawater-cooling system for the various industrial processes. Stage I of this provided a flow of 113 m^3/s – equivalent to the mean annual discharge into the Gulf from the Tigris and Euphrates rivers – with provision for doubling this later.

Sir Alexander Gibb & Partners was appointed to carry out a feasibility study in 1978 and went on to complete the detailed design. The choice of sites for the intake and outfall was critical because of the need to avoid recirculation with warmer water finding its way back into the intake. For environmental reasons, heat dissipation studies were carried out to ensure acceptable levels of the ensuing temperature rise in the Gulf waters. Gibb carried out extensive modelling to optimise the system. As currents in the Gulf flow generally counter-clockwise, i.e. north to south at Jubail, it was decided to site the intake on the north side of the industrial causeway and the outfall on the south within the industrial harbour.

A 4-m channel dredged alongside the causeway and protected by a rock bund allows water to be drawn from a deeper level where it is cooler. A stainless steel bar screen and seven 12-m diameter drum

above

Saddam International Airport, Baghdad, was built in the 1970s but designed to be
extended to cope with expected traffic to the year 2000.

© G. Maunsell and Partners

above

Baghdad's elegant Adhamiyah bridge is sited on a curve of the Tigris.
The cable-stayed design avoids the need for a pier on the outside of the curve,
where scour would have been a problem.

© G. Maunsell and Partners

left
Riyadh's conference centre is considered one of the finest buildings in the Kingdom.
© Ove Arup & Partners

below
The sports hall at King Abdul-Aziz University in Jeddah was the world's largest totally enclosed cable net tent structure when it was built in the late 1970s.
© Try Construction Group

above
Buildings such as the
Kuwait Foundation for the
Advancement of Science
demonstrated that high
standards of construction
could be achieved in the
Middle East.
© Ove Arup & Partners

left
Detail from the
Kuwait Foundation
© Ove Arup & Partners

right and below
The port and industrial
area at Umm Said were
built as part of a
drive to diversify
Qatar's economy.
© *Sir Alexander Gibb*
& Partners

above
Dubai Dry Dock epitomises the enormous scale of some Gulf projects.
Four years were allowed to construct three docks, the largest of which
could hold a 1 million-dwt tanker. No ship of this size progressed
beyond the drawing board.
© Costain Group

top
Al-Maqam horse racing grandstand, Al Ain, in which a series of concrete arches support precast post-tensioned roof beams cantilevering 25 m.

right
Abu Dhabi National Day Parade grandstand called for post-tensioned floors and beams, which in turn needed concrete stronger than had ever been produced in the UAE before.

above
Al Ain football stadium uses a similar structure to the al-Maqam grandstand. The roof and supporting arches are a completely independent structure from the seating.
all photographs © Jan Bobrowski and Partners

above

Abu Dhabi Sheraton hotel was built using fast-track methods as part of a
crash programme to ease the Emirate's shortage of hotel space.
© Sir Alexander Gibb & Partners

screens guard the intake from the ingress of debris and marine life. A pumping station is sited on the shore where 14 axial flow pumps, each of 10.3 m³/s capacity, pump the water into a supply channel to the supply compartment of the cooling system's main canal. From here gravity offtakes supply cooling water to sumps on each industrial site. After use it is pumped back into the return compartment of the main canal where it flows by gravity to the discharge headworks.

The main canal is approximately 13 km long in three discrete lengths serving different parts of the industrial complex, the lengths being connected by inverted siphons. The canal cross section is divided into three compartments; one for supply and one for return, with the central section capable of being used for either supply or return when one of the other sections is out of commission for maintenance. The supply channel connects the pumphouse to the main canal and a return channel connects the discharge headworks at the downstream end of the return compartment to a cascade spillway into the harbour. Water leaves the system up to 10°C warmer than it entered but this will not adversely affect marine life, says Gibb director Tim Woods Ballard. Gibb later acted as the designer for the cathodic protection system which was installed to protect the system from concrete deterioration.

On the other side of the country, about 300 km north of Jeddah, attention focused on the small trading port of Yanbu. This was the traditional entry point for pilgrims *en route* to Medina and was a flourishing trading harbour for small vessels. Before expansion it had just two berths along an L-shaped quay, 170 m and 210 m long respectively.

In the second five-year plan Yanbu was selected with Jubail for industrial development. Under the joint scheme, two pipelines were to be constructed across the Arabian peninsula, one to deliver up to 670,000 b/d of crude oil for export to Europe and the US, and the other to supply natural gas, which had previously been flared off, for use in a new petrochemicals' complex to be built adjacent to the port.

Halcrow again appointed consultant for the creation of a modern deepwater harbour at Yanbu. Again, the number of new general cargo berths rose from two in 1976, to seven, dredged to a depth of 11 m or 12 m. Two of the berths were allocated to handle bulk cement at a throughput of 700,000 t/y. Mechanised handling equipment was used to transfer cement into a 20,000-t silo at what at the time was believed to be the biggest bulk cement terminal in the world.

Water supply

Ensuring adequate water supply is crucial in Saudi Arabia as every-where in the Gulf. J. D. & D. M. Watson was involved from 1966 on supply to the Holy City of Mecca. As non-Muslims are not permitted there, only Muslims could carry out site-supervision. Watson used closed-circuit television, with Muslims operating the cameras and non-Muslims sitting in air-conditioned comfort in their offices outside the wall, to assist in supervision of construction.

US$110 million has been spent to date on Mecca's water supply. Balfours (now part of Maunsell), Binnie & Partners, Watson and John Taylor & Sons have all designed water or wastewater schemes around the country. But the biggest challenge was the Riyadh water-supply project.

In the early 1970s Riyadh was developing at a furious pace. When in 1824 Ibn Saud designated Riyadh as his capital, it was a small desert village surrounded by palm trees. The city was probably chosen as the capital because of its central location in the Kingdom and the Arabian peninsula and because of its water supply from underground aquifers.

At the beginning of this century, Riyadh occupied a circular area roughly 0.75 km in diameter. For many years after World War II it was almost a closed city, and it was hard for foreigners to get in. Jeddah was the main commercial city, where all the foreign embassies were located, as well as the main gateway for pilgrims visiting Mecca. Riyadh grew only slowly up to 1955 when ministries and government agencies were transferred there. A railway to Dammam was completed in 1950 and the first airport was built in 1952. By the 1950s Riyadh covered 8.5 sq km. This increased tenfold in the next 20 years. A formal planning depart-ment was set up in the late 1960s and a master plan foresaw the city increasing to 300 sq km within 20 years. In fact, by the mid-1980s, an area of 1,600 sq km had already been covered.

In these circumstances, it is not surprising that provision of water was a major consideration. Sir M. MacDonald & Partners was appointed by the Ministry of Agriculture & Water in 1972 to identify a site for a new wellfield to develop an additional supply of water. Shallow aquifers and nearby wadis had been fully exploited early in Riyadh's history and groundwater from the deeper Minjur aquifer had become increasingly important as a source of supply.

Originally, says Mott MacDonald's water-supply and wastewater director Mike Burley, the plan was to develop that source further. As part of the same study the consultant investigated the deeper Wasia aquifer, which slopes away from Riyadh towards the east coast. Fifteen wells and ten piezometers were drilled at depths ranging from 350 m to 800 m in the Wasia and from 200 m to 660 m in the Minjur. It was concluded that potential for further development of the Minjur was limited. The Wasia, on the other hand, could provide water of a reasonable quality 'in vast quantities'.

Total dissolved solid content of the Wasia was about 1,200 ppm. Though this was better than existing supplies to Riyadh, it was near the upper limit recommended by international guidelines. The feasibility report suggested that this could be used directly; alternatively it could be desalinated.

The ministry accepted the report and commissioned Sir M. Macdonald to go ahead with detailed design, but it decided that the water would be desalinated, and that the capacity of the scheme would be doubled to 200,000 m^3/d. Construction of the world's biggest desalination plant was envisaged. Unlike the multi-flash plants serving the coastal cities, this was to be a reverse osmosis plant.

Preliminary designs in the feasibility report looked at wellfield sites and pipeline routes to the capital. The chosen wellfield was 110 km from the capital just north of the Riyadh–Dammam road. It was to consist of 62 wells, an average of 400 m deep though some went down 750 m. The wells were 400 mm in diameter with pumps at 300 m depth. A treatment plant, power and pumping station complex were to be sited about 20 km from the wellfield.

Because of the depth from which the water is pumped it reaches the surface at a temperature of 56°C and is first cooled. A softening plant is next provided – originally this was intended to prevent formation of scale on the membrane of the reverse osmosis plant. Lime and soda ash are added to precipitate calcium carbonate and magnesium hydroxide: this is deposited as sludge in tanks as the water flows upwards, with a rapid gravity sand filter to remove the last traces of precipitant. Two further tanks thicken the sludge, with supernatant water returning to the inlet. Around 100 t of lime is used daily, and sophisticated chemical handling facilities are provided to deal with this.

A highly complex automated monitoring system was provided for the treatment works, with instrumentation developed by Mott, Hay & Anderson (with which Sir M. MacDonald was later to merge). Telemetry connects all 62 wells to the central control room. Any fault sounds an alarm and alerts maintenance staff. The system 'zeroes in' on the fault, identifies what has gone wrong, informs the maintenance department what spare parts are needed and automatically re-orders them. 'It's probably the most sophisticated monitoring system we've ever done,' says Burley, who admitted to having had worries over the reliability of such facilities. DEC was chosen as computer supplier because it had a 24-hour maintenance service in Riyadh and so could provide a rapid response to any breakdown. But the worries proved unfounded.

Buildings at the treatment site have unusual space-frame double-skin roofs to allow air to circulate, causing cooling by natural convection and eliminating the need for air-conditioning.

The reverse osmosis plant originally planned was never built. Instead, it was decided to pump distilled water from a multi-flash evaporation plant, built independently by the Japanese on the coast. This would be mixed with the Riyadh water.

To convey the water from the Wasia treatment works to Riyadh, there is a pump station at the treatment works with six 43,000-m³/d capacity 1.1-MW pumps and two 21,500-m³/d 0.6-MW pumps. A booster pump station part way along the pipeline route has identical capacity; about two-thirds along its length the pipeline reaches a high point where there is a break-pressure tank from where water gravitates the rest of the way. Water from the distillation plant is also piped to this high point for mixing. At Riyadh the water is received in a 150,000-m³ concrete reservoir which contains 14 hours' supply for the city.

The pipeline consists of twin 1,000-mm diameter pipes of two types: ductile iron-lined with cement mortar, and steel-cyclinder-reinforced concrete pipes. Ground resistivity was measured at 400-m intervals along the route to check for potentially corrosive ground conditions. 'To our great surprise we found some areas of very low resistivity,' says Burley, 'yet the ground looked the same as everywhere else. We found this hard to believe.' But an exploratory excavation revealed a yellow water-bearing clay at the low resistivity sites. 'We had to put in cathodic protection for selected lengths,' he says.

Two additional reservoirs have been added more recently at the high point, to give additional security of supply. Built in reinforced concrete, they each have a capacity of 1.5 million m³ and cover an area of 725 m x 325 m.

Because of the isolation of the treatment works, a village was constructed for the people who work there and their families – a total of 400–500 people. This included 248 houses, 186 flats, two schools, a mosque, shops, and a sewage works.

Agriculture

The Wasia scheme was a ground-breaking project for Sir M. MacDonald: the initial study was the firm's first job in Saudi Arabia, and it led to a further commission – the £23 million Umm er Radhuma study.

Groundwater Development Consultants, a subsidiary of Sir M. MacDonald, was appointed to do the study in 1977. The Umm er Radhuma is one of a series of limestone aquifers which outcrop in the west of Saudi Arabia towards the Red Sea hills and slope down to the east. Water enters these aquifers at the outcrop and migrates down the slope more or less in balance with the loss of water from the other parts of the aquifer.

The geology is such that the same aquifers exist below Iraq in the north and Yemen and Oman in the south. Age testing of the water samples indicate that it may have taken 20,000 to 30,000 years for a molecule of water to flow down the system from the outcrop to an abstraction point about 800 km away.

The area to be studied for agricultural development was the whole of the Eastern Region of Saudi Arabia, a total of 35 million ha.

The study revealed that the area available for large scale irrigated agriculture was substantially less than had been expected on the basis of an earlier survey. The study also showed that the best soil did not coincide with the areas where water quality was best. Good soil was scattered and discontinuous. The result was that whereas the initial brief had been to select the best 50,000 ha of land from a supposed 1.5 million ha, in the end it was only possible to identify 11 'priority areas' totalling 55,000 ha which, when irrigated, corresponded to a net area of 34, 240 ha. All the water was high in sulphate content.

Centre pivot and linear movement sprinkler irrigation systems were recommended in farm units with areas between 20 and 100 ha. Much attention has to be paid to maintenance because the precipitate from the high sulphate water tends to clog the nozzles quickly.

However, crops were in the end more successful than expected. It was discovered that high wheat yields could be obtained, although significant amounts of fertiliser were needed. Fine sand proved to be the most successful soil type. With the encouragement of generous government subsidies, there was substantial private development and the country quickly achieved a wheat surplus. This has remained despite drastic cutbacks in the level of subsidy.

The Saudi Arabian government also paid for the study to be extended to Bahrain, which is also underlain by the Umm er Radhuma aquifer.

But neither the soil nor the water quality showed much promise.

David Donald, the project manager for the study from 1977–80, believes that, for Saudi Arabia itself, the results were well worth while, demonstrating that it could become self-sufficient in food, and providing a basis for future development towards that end.

Others are more pessimistic. As one expert puts it: 'Saudi Arabia is drawing down its natural water resources by the development of agriculture within the country. A potential crisis looms.

'The crops they are producing, such as wheat, are not economically viable. They need to control water use far more rigidly and adopt a more realistic approach to reusing water on a widespread scale.'

Agriculture may need to develop away from intensive mechanical methods to more natural community-based methods for sustainability, he suggests. 'This could involve using effluents for afforestation and green belts to give shade and create microclimates to enable them to grow other sustainable crops.'

Drainage and sewerage

Both Watson and John Taylor & Sons were responsible for extensive drainage and sewerage schemes throughout the Kingdom.

Watson was engaged on a major scheme in Jeddah, where severe flooding was a problem, aggravated by run-off from the foothills of the Hejaz mountains which attempted to follow the routes of the old wadis which crossed the city. The firm was commissioned in 1967 to report on the problem as a result of which stormwater protection was carried out in two phases.

In phase I two stormwater protection channels totalling 29 km in length were built, one to the north and one to the south of the city. The open channels intercept flood water and divert it around Jeddah into the Red Sea. The Mashwab dam contains flood waters and regulates flow into the northern channel, while the southern channel accepts flow directly. Several road bridges across the channel were also included in the scheme.

An in-town drainage system made up phase II, which began in 1976. A network of 65 km of large-diameter pipes and reinforced concrete culverts was constructed to drain flood-susceptible areas within the city. Mostly, this drains by gravity to the Red Sea. But for the commercial centre, which is on the low-lying coastal strip, stormwater had to be pumped.

In 1975, a major health hazard became apparent in Jeddah: large ponds of foul water were forming at ground level. This was caused by rising groundwater levels from increasing septic tank discharges, mainly because of higher consumption as a result of improved domestic water supplies. Watson was instructed to prepare designs to deal with the problem urgently.

An emergency solution was developed which involved constructing 13 km of open-jointed pipes surrounded by granular material and connected to the existing trunk sewer system. This eliminated the hazard and protected many buildings with shallow foundations which would have been threatened by rising damp. In 1981 this interim solution was superseded by installation of a network of minor sewers and house connections. Watson also produced feasibility studies for effluent reuse in Jeddah and Mecca in the early 1970s.

Between 1974 and 1982 John Taylor & Sons solved a similar problem in the three oasis cities of al-Hasa, Buraidah and al-Qatif. As in Jeddah, no main drainage facilities other than septic tanks existed, and the firm designed foul and stormwater sewerage systems. Stormwater was discharged to irrigation drainage channels or large soakaway areas while the foul sewers terminated at treatment plants. In all, 310 km of foul sewers, 75 km of surface water sewers, 30 pumping stations and three treatment works comprising waste stabilisation pond systems were built to serve a combined population of 200,000. At Buraidah a 6-km flood-diversion channel was also included.

Effluent was treated to a suitable standard to allow it to be reused in the irrigation of date palms – the first time this had been done in Saudi Arabia, although Watson had produced feasibility studies for effluent reuse in Jeddah and Mecca in the mid-1970s.

In 1982 a treatment works designed by John Taylor & Sons as part of a £500 million scheme, including both water-supply and sewerage, included facilities to bring the effluent to WHO standards for potable water. Treatment included rapid gravity sand filtration and chlorination, ozonation and carbon filtration. The plant was designed to serve a population of 500,000. This non-potable water was to be distributed around the city in a separate distribution system.

It was not until the mid-1980s, however, that an effluent reuse scheme, supervised by Taylor, came into effect at the Taif sewage treatment plant, where a water-reclamation plant was designed and built to provide a flow of 100,000 m^3/d to potable water standards. Sewage effluent was treated with alum and lime in mixing basins, followed by flocculation, flotation, and multi-media filtration. It was pumped to storage reservoirs of 60,000 m³ capacity where it was chlorinated before passing through activated carbon filters to absorb organic matter and chlorinated hydrocarbons, before distribution through a non-potable distribution network.

During the same period Taylor also prepared studies and designs for effluent reuse for agricultural purposes in the Qassim Region, within a 50-km radius of Buraidah and Unayzah, and in Abha the capital of the Azir Region.

It is interesting to note the contrasting approaches of three different states which used effluent reuse extensively. Abu Dhabi used the water for municipal beautification – the irrigation of verges, roundabouts and central reservations. This had a dramatic effect on local ecology as flowers and other vegetation attracted insects. Dubai adopted a similar policy, as did Qatar where the irrigated Doha racecourse was a notable success. The only restriction was that, in public places, irrigation had to be carried out at night because of the possible danger to the public from airborne spray from sprinklers.

By contrast, Kuwait reused its effluent to grow animal fodder such as alfalfa. Saudi Arabia, meanwhile, piped the recycled wastewater back to households via a separate distribution network, distinctively marked to reduce the risk of inadvertent cross-connections.

Buildings

Construction of highly prestigious buildings was naturally popular during the period of rapid expansion and British consultant Ove Arup & Partners was responsible for a considerable number. Two of the earliest were hotel and conference centre complexes in Riyadh and Mecca.

Ove Arup's involvement in Saudi Arabia dates from 1966 when the firm was invited by the architect Trevor Dannatt to collaborate on a design for a competition for the Riyadh complex. The two had earlier collaborated on a steel bridge design. The Saudi Arabian government had realised the need for a conference centre after an attempt to hold an international conference in Riyadh had demonstrated the inadequacy of existing facilities. Dannatt's design won, and early in 1967 representatives were summoned to the Kingdom to discuss the project.

Though the competition had only mentioned Riyadh, it became apparent that the intention was to build three complexes, the others being in Mecca and Dhahran. While in Saudi Arabia, Ove Arup's Ted Happold made contact with the winners of the design for the Mecca project, architects Rolf Gutbrod and Frei Otto, the latter famed for his designs of cable and tent structures. As a result, Ove Arup was appointed to be engineer on the Mecca project also.

Both projects went out to tender in 1968. COGECO of Italy won the Riyadh project while the Mecca complex went to a joint venture of Entreprise Thinet of France and Joseph Khoury of Lebanon.

Of the two, Mecca is the more structurally complex. It is situated next to the Mecca–Jeddah highway, 7 km from the *Haram* (the central mosque). The site is at the junction of several valleys and is prone to flooding. A riprap embankment was formed to protect the site and the general site level was raised 1.5 m.

The project is a complex of low-rise buildings. A 200-bedroom hotel consists of three- and four-storey blocks, grouped round an artificial oasis. The conference centre includes a 1,500-seat main auditorium 21 m high, with two-storey foyers containing three seminar rooms seating 200, around another oasis. There is also a mosque.

The hotel is of reinforced concrete slabs spanning onto cross walls. The ground consisted of fine, compacted sand so reinforced concrete pad foundations could be used. The hotel's floors are arranged as a series of terraces stepping back from the oasis. At ground level, where a large open space was needed for the restaurant, the cross walls above are supported by Y-shaped columns, with the walls spanning as deep beams between columns.

The auditorium is built of precast seating steps supported on portal frames radiating from the stage. Its most striking feature is the roof, a cable-suspended system independent of the main structure. The front edge is supported by a portal frame formed by a steel beam and two raking masts. The rear edge is supported by three boundary cables connected to six raking masts. Steel cables spanning between the portal frame and the boundary cables form the roof itself, which is curved in a single plane. These cables support T-section steel members carrying the roof cladding, sandwich panels of insulation and aluminium sheeting.

A similar roof spans the foyer area, but its cables are supported at both ends by boundary cables anchored to the cross walls of the seminar building. Vast canopies of open timber lattice-work act as sunbreaks over the oases, and these are also supported by cables and three or four masts. For the canopies, geometry of the cables and the loads they carried were interdependent, so the structure had to be analysed by making an estimate of the geometry and using a computer to do an iterative analysis until the loads and geometry became compatible.

This was naturally a highly complex structure to erect. Professor Rolf Gutbrod's philosophy was that close collaboration with the contractor was essential if such structures were to be built successfully. According to Ove Arup's house magazine, *The Arup Journal*, in September 1971, the project team had a hard time convincing the contractor of this. It was not until a trial erection was carried out at the works of the steel supplier, Voyer, that the potential problems became apparent and a more productive dialogue ensued. 'It was only after the first attempt at erecting the cables had to be abandoned, owing to some most visible inadequacies, that our dialogue improved,' said the *Journal*.[5]

Trial erection allowed a decision to be made on the method of final erection. The designer wanted to prefabricate large sections of the roof to minimise construction on site, which would, however, have made transport more difficult. The contractor preferred to erect everything item by item on site. The trial proved that both methods would work, and in the end the contractor stuck to the non-prefabricated approach.

Another major source of frustration was site supervision, since the site was in the Holy City and therefore non-Muslims were forbidden to enter. Initially the contractor tried to supervise the work from outside the forbidden zone, but this proved impossible. Instead it recruited several French Muslims at levels from project manager to trade foreman. This system was generally satisfactory although work was thrown into chaos on one occasion when entry was arbitrarily denied to the supervisors.

Structurally less innovative, the Riyadh complex was nevertheless considered one of the finest buildings in the Kingdom. It is sited to the west of the city, in what at the time was an undeveloped area near the new airport. Again, the main buildings are the hotel and conference centre, but the complex also includes a mosque, villas for hotel staff and service buildings. The conference centre includes a 1,400-seat auditorium, with a foyer on three levels providing space for exhibitions, plus five committee rooms. The hotel has 200 bedrooms and is W-shaped in plan and has six storeys.

Construction is generally in reinforced concrete with the concrete surfaces – whether smooth, boarded, with exposed aggregate or bush hammered – forming the finished surface. As with the Mecca complex, a good deal of thought had to go into the provision of expansion joints to cope with temperatures rising to 50°C.

The hotel is also of concrete cross-wall construction set out on a triangular grid, which made setting out of the walls on plan highly complex. The auditorium is enclosed by 250-mm concrete walls with piers of a total thickness of 450 mm. Seating steps are formed in in-situ concrete spanning on to concrete beams or longitudinal walls. Its roof is a two-layer space-frame on a 4.33-m square grid, 3 m deep and built from Grade 52 steel. It is 56 m square overall, supported by four columns at the corner of a 39 m square. Because of the closeness of the airport, a 100-mm concrete slab covers the top of the space-frame to provide sound insulation.

All the space-frame members are made up from angle-section steel arranged in twos or fours. At each node four horizontal and four diagonal members are connected by bolts to 20-mm or 30-mm steel plates. There are 365 nodes and 30,000 bolts.

Ove Arup was also overall consultant for King Abdul-Aziz University in Jeddah. The university's sports' hall was remarkable for using what was then the largest totally enclosed cable net tent in the world, covering 8,500 sq m. Like the Mecca conference centre, the £6 million project was designed by architects Rolf Gutbrod and Frei Otto. Consultant for the superstructure was Buro Happold, the firm set up by former Ove Arup engineer Ted Happold.

The contractor was W. S. Try (International) working on its first overseas project, which began in early 1978. Though such an unusual structure might be thought undesirable for a newcomer to overseas work, it did not pose problems, explained Hugh Try and Michael Rush in a paper to an ICE conference on 'Management of International Construction Projects' in 1984. 'The superstructure, although unusual, was structurally very simple in concept and was to be supplied and erected very largely by nominated subcontractors. The project was of about the right size from the bonding and interim finance point of view and our own work in the foundations and floor slab appeared reasonably uncomplicated.'[6]

Foundation design was Ove Arup's responsibility. A 500-mm deep reinforced concrete ring beam pinned down by 10-m long ground anchors at 3-m centres – a total of 187 – ran round the perimeter of a 150-mm thick reinforced concrete slab. Along the beam 24 massive reinforced concrete anchor blocks, 1,500 mm deep and each held down by seven ground anchors, provided anchorages for the main cables of the tent structure.

There was some disagreement among geotechnical experts about whether ground anchors were in fact capable of resisting the forces which would be applied to them. However, tests on installed anchors showed that they were completely safe.

The superstructure consisted of an 80-t net supported by eight 30-m high steel masts of 800-mm and 600-mm diameter. Subcontractor for the superstructure work was the Swiss company Habegger, whose specialism was the construction of Alpine cable-car systems. Erection of the tent roof called for similar techniques. Habegger appointed the German consultant Ingenieur Planung Leichtbau to provide a cutting pattern with 3-mm tolerance for all the cables, which were supplied in lengths ranging from 2 m to 145 m.

Prior to erection the cables which were to form the net were laid out on the ground around the masts to a prearranged pattern on an approximate 500-mm grid. At each intersection the cables were secured together by a clamp. Laying out the cables and fitting the 57,000 clamps took eight weeks. The net was lifted into place in two stages. First it was hauled upwards from eight points simultaneously by pulleys clamped to the mast heads. When two-thirds up, the net was fixed to the ring beam edge clamps. This took about two days. 'More man-hours were spent on the design of the erection than on the actual erection itself,' Ted Happold told *New Civil Engineer* at the time.[7]

For the second-stage lift, the pulleys were removed from the masts and temporary steel beams were fitted to the bottom of each mast to allow the masts to be jacked upwards around 400 mm, tightening the net and giving it its final shape. This was a crucial stage in the operation as it largely determined which of the wrinkles in the net from the first stage remained in the final structure, though small inconsistencies could be removed by adjusting grid clamps. Final prestressing had to be done early in the morning when the temperature was around 37°C.

The cable net supports an outer membrane of white translucent acrylic-coated material encased with PVC. An inner membrane is suspended from hangers below the cable net and is lighter and more translucent than the outer material. The outer coat, originally white, soon gained a coating of sand giving it a beige colour, but even so enough natural light continues to penetrate for most daytime activities. The plenum between the membranes is designed to act as a large ventilation duct for the tent, from which fresh air for the air-conditioning system is drawn.

Tight tolerances were called for in the construction of the base slab and W. S. Try had hoped to be able to erect the tent before casting the slab, in order to make it easier to protect it from the heat of the sun, and the associated risk of shrinkage and cracking. But delays in detailing and fabricating the superstructure prevented this and the slab was cast first. In the event, said Try, 'this worked extremely successfully and demonstrated what high quality work can be achieved by careful supervision and using relatively unskilled imported labour from the Third World.'

Reinforced concrete supports for the seating, and precast terrace seating units, were cast on the sports hall floor and lifted into place before fixing the internal membrane.

A tent roof was also used for Riyadh's 66,000-seat King Fahd international stadium, built between 1982 and 1986. The designer, the Ian Fraser, John Roberts partnership, was given the brief of creating something unique and providing a clear line of sight for all the spectators. This meant providing a roof with a clear span of 290 m. PTFE-coated glass-fibre fabric was the solution.

The roof was designed and built by US specialist Birdair. Erection had to be carried out at night when convection currents caused by the sun had died away. In dazzling white fabric, the roof has 24 peaks supported on hollow steel columns. It was erected as 96 panels, each 800 m^2 in area and weighing 4 t, including fittings.

The main structure of the stadium is relatively conventional and was built by Bovis International and West Germany's Philipp Holzmann. It comprises a concrete grandstand on one side and a lavish, five-storey royal pavilion on the other. The pavilion is equipped with bedrooms, staterooms, dining halls and a 76-seat area reserved for the royal family and guests.

Roads

Shifting sand dunes caused problems for Ove Arup on the firm's biggest project in Saudi Arabia, the Trans-Saudi Expressway. The consultant was appointed in 1976 to design 670 km of the six-lane dual carriageway expressway connecting Jeddah, Riyadh and Dammam, much of it through desert and oilfields. The road was designed to full motorway standards and included 44 intersections and 57 bridges.

A problem on all roads in the deserts of Saudi Arabia is the drifting of wind-blown sand on to the road surface. Only a few centimetres is needed to constitute a driving hazard with the high vehicle speeds which are common.

Ove Arup studied the problem and solved it by three aspects of its road design. First, four types of desert terrain were identified as being particularly susceptible to the problem and so the route was designed to avoid those where possible.

Second, the firm concluded that features of the road design itself could accentuate the problem. Typically the sand was blown and bounced safely across the surface of the road by a process known as saltation, provided the wind speed was not slowed by any obstruction. But at crash barriers, cuttings and bridge abutments, 'wind shadow' could lead to the deposition of sand. The aim therefore was to streamline the cross sectional shape of the road so that, just as with an aeroplane wing, the air and the sand it contained streamed freely across the full width to the far side.

Third, the sand surface alongside the road was stabilised to prevent sand creeping on to the road. This was achieved by using an oil or oil-extended latex to stabilise the sand. A depth of penetration of 100 mm was specified to guard against surface erosion. It was found most efficient to apply the stabiliser in up to four strips 2 m wide, with 4 m gaps between them, on each side of the road.

Saudi Arabia today

Today, Saudi Arabia remains a difficult, highly competitive market. Even in the late 1970s local contractors were becoming capable of handling £10 million to £20 million jobs and local consultants have also gained considerable expertise. Some firms still see the market as having possibilities – in mid-1993, for example, Wimpey was finalising registration of a new company, Wimpey Saudi. But many others see little possibility of winning work at competitive rates. Ove Arup, for example, remains in the market only to provide specialist skills such as fire-engineering, geotechnics, or tall building design which local firms cannot provide. 'For the average building project it's not worth our while bidding, because we cannot deploy our skills properly for the fees on offer.' says Ove Arup's Middle East projects manager John Ward. 'We are keen to work there, but most British firms find it tough.'

chapter six

KUWAIT

————

Saddam Hussein's invasion of Kuwait in 1990, and his attempt to incorporate it as Iraq's nineteenth province, was not entirely without precedent. As soon as Kuwait became independent of Britain in 1961, the then dictator of Iraq, Abd al-Karim Qasim, threatened to occupy the embryonic state. The threat was faced down with British and then Arab League help.

Kuwait's size has been a source both of strength and weakness. Its area is a mere 17,818 sq km, and its population in 1985 was only 1.7 million. Before the 1990-91 Gulf War only 40 per cent of the inhabitants were Kuwaiti citizens, the rest being immigrant workers, mostly Palestinians and Egyptians. Since then the population has declined as many Palestinians were expelled. Immediately after independence, there was doubt about whether the state was big enough to be viable. On the other hand, the extent of its oil reserves in relation to its population allowed its people to attain high standards of wealth quickly, in contrast with more populous countries like Iran. Kuwait became, in fact, the Gulf's first oil-producing 'city state', and provided the model for Abu Dhabi and Dubai several years later.

Although some dismissed the state as an artificial British creation, Kuwait can lay claim to an independent existence dating back to 1710, when the Utab tribe from north and central Arabia took control. Until then it had been an unremarkable fishing village which had not been able to contribute to trade in the Gulf which centred on Basra in Iraq. Nominally Kuwait was under Ottoman control, but provided this sovereignty was recognised Istanbul allowed the Utab independence of action. The Utab developed Kuwait as a terminus for trade from Syria and the Levant.

In 1760 there was a split in the ruling Utab tribe, with one faction, al-Khalifah, emigrating to Bahrain. The faction that remained, al-Sabah, created the emirate of Kuwait.

The British East India Company set up a base in the Gulf in 1776, recognising the Gulf's importance in control of access to India from Europe. (Indeed, the Indian rupee became Kuwait's official currency and remained so until independence in 1962.) Britain became interested in Kuwait itself only towards the end of the nineteenth century when there were suggestions that Russia wanted to establish a coaling station in the Gulf, and also that German interests might seek to extend the proposed Berlin–Baghdad railway to Kuwait because it could offer better harbour facilities than Basra.

Instead, Kuwait suggested, and Britain accepted, that the amir should receive an annual sum from Britain in return for not granting concessions over any part of his territory. Eventually, in return for further economic concessions, Kuwait formally became a British protectorate in 1914.

Pearl-fishing had been a mainstay of the economy but declined in importance in the 1930s. At about the same time negotiations on an oil concession began between Amir Ahmad and the Anglo-Persian Oil Company (APOC, now BP) and the US company Gulf Oil. With the world depression curtailing investment, there was little sense of urgency about the talks. Britain had, as before, initially insisted on excluding non-British companies. Eventually, however, as in Bahrain, it was persuaded to modify this stance.

Discovery of oil and the granting of a concession in Saudi Arabia lent a new urgency to the talks. APOC and Gulf Oil agreed to form a 50/50 alliance, thereby establishing the Kuwait Oil Company (KOC) which, in 1934, was given a 75-year concession covering all Kuwait. Four years later a discovery was made in the Burgan field, the world's largest at the time. Production, delayed by World War II, eventually began in 1946.

Oil revenue increased rapidly: from £3 million in 1949 to £60 million in 1952. Amir Abdullah Salem al-Sabah envisaged western standards of living within a generation.

During the 1950s, Kuwait remained, to all political intents and purposes, a British colony. One obstacle to full independence was considered to be the shortage of educated Kuwaitis who could administer an independent state. However, at this time, oil wealth drew in large numbers of Palestinian, and to a lesser extent, Egyptian, workers. This immigration led to a rise in Arab nationalism and demands for a change in the relationship with Britain.

Independence was granted in June 1961. Iraq's dictator at the time, Qasim, immediately threatened to occupy Kuwait on the grounds that it was formerly part of the Ottoman province of Basra. At the request of the amir British troops were sent in. But in a shrewd move, they were replaced two months later by an Arab League force, demonstrating recognition of Kuwait as an independent state by the rest of the Arab world. Kuwait emphasised this independence by formalising diplomatic relations with the Soviet Union and taking a non-aligned stance with the Cold War powers.

It was also the first state in the region to have an elected assembly. The 50-member assembly was first elected in January 1963 by the 10 per cent of the population who were eligible to vote. Non-Kuwaitis had no vote. The assembly's power was limited but it became an effective voice for criticism of policy. Elections were held at four-yearly intervals until 1976, when Amir Jabir al-Sabah was sufficiently embarrassed by statements in the assembly to dissolve it. It was restored five years later in the hope that it would help to guarantee stability, after the Iranian revolution was welcomed by Kuwait's Shiite minority and its large number of Palestinian workers. It was dissolved again in 1986 but in response to popular demand elections were held again in October 1992, producing an assembly with an opposition majority.

Oil remains the basis of Kuwait's economy, accounting for 50 per cent of the gross domestic product and 90 per cent of exports. The oil industry was nationalised in 1975. To reap the maximum benefit from its oil Kuwait built three refineries, now being upgraded in the aftermath of the Gulf War, and petrochemicals and fertiliser plants. Fishing is now relatively unimportant; agriculture was insignificant but, through irrigation, Kuwait has come within reach of self-sufficiency in fruit and vegetables.

Some industries have been established including a local construction industry which, however, depends heavily on government expenditure for its work. Attempts to develop a financial services sector have had some success; Kuwait was the first Gulf state to have a stock exchange.

Foreign investment overseas by the government has proved a major source of income, earning almost as much as oil exports. Kuwait was also the first Middle East oil producer to buy an oil distribution network in Europe and an oil engineering company in the US in efforts to create a worldwide integrated industry.

Kuwait became one of the world's most generous aid donors, with aid being provided directly by the government and through the Kuwait Fund for Arab Economic Development (KFAED), which gives grants and long-term low-interest loans for development projects in other Arab states.

The question of Kuwait's continued independent existence was brought dramatically to world attention in August 1990. Saddam Hussein had threatened the state because of its reluctance to allow Iraq secure access across Kuwaiti territory to the Gulf, and because Kuwait was producing oil at a rate above its OPEC quota, which Iraq claimed was costing it a billion dollars a month in lost revenue. Saddam Hussein also claimed that Kuwait was abstracting oil which he regarded as belonging to Iraq. Saudi Arabia attempted unsuccessfully to mediate, but few expected Saddam Hussein's invasion on 2 August. Kuwait's annexation as Iraq's nineteenth province was announced.

Saddam Hussein had failed to anticipate world opposition to the move, particularly when he also moved troops towards the Saudi Arabian border. Had the Iraqi leader also gained control of Saudi Arabian oilfields as well as those of Kuwait, he would, with Iraq's own oilfields, have over half the world's oil supplies in his power. Accordingly the US responded quickly to requests for support from Kuwait and Saudi Arabia, and over the ensuing weeks a 600,000-strong contingent of land, sea and air forces was built up. Although US, British and French troops made up the bulk of the force, some 54 countries were represented in it; the US was at pains to show that this was not a western crusade against Islam.

Sanctions against Iraq imposed by the United Nations Security Council were highly effective but failed to move Iraq. In the end, after Saddam Hussein failed to withdraw his troops by a United Nations approved deadline of 15 January 1991, an attack against Iraq was launched. Five weeks' bombing of Iraqi targets and Iraqi forces in Kuwait commenced, to be followed by a ground assault. Kuwait was liberated, but at the cost of damage to 90 per cent of the state's oil infrastructure by retreating Iraqi troops.

Burning oil wells were capped by several US and British teams in an operation which lasted seven months. Considerable ecological damage was caused by the smoke, and by oil deliberately spilled into the Gulf, though this was less than some analysts predicted. Kuwait is now taking the opportunity to replace much of its oil production infrastructure, which was in any case nearing the end of its life.

The fortunes of British consultants and contractors in Kuwait have been mixed. From World War II to the first few years after independence they were reasonably successful, but in the 1960s, and during the peak of activity in the 1970s, British firms had fallen from favour considerably. An idea had spread that some firms had taken advantage of the Kuwaitis, and in the early 1960s many left. In the 1970s some managed to re-establish themselves, but it was a struggle.

Contractors found the going particularly hard. A development plan began in 1952 and, on British insistence, five British contractors – Costain, D. & C. and William Press, Taylor Woodrow, John Howard & Co and Holland, Hannen & Cubitts – were given all the work on a 'cost-plus' basis of 15 per cent of total expenditure. This caused considerable resentment among Kuwaitis and in 1954 the monopoly was ended and competition from other countries allowed.

In 1964 Kuwait adopted fixed-price contracts and brought in stringent conditions which did not comply with normal rules for international tendering. Claims ceased to exist. Tender law was also rewritten, with strict rules on who should receive and decide tenders. There were also tougher requirements for performance bonds which could be called in at any time, though this option was rarely, if ever, used. Forms used on site to issue variation orders had a space for extensions of time to be filled in – with the word 'nil' pre-printed.

These new onerous conditions were enough to put most British contractors off: the British were replaced by Dutch, Japanese and American firms, some of whom predictably fell foul of the contract regime. By the late 1970s, however, as fewer international contractors were willing to bid for Kuwaiti projects, conditions were reported to have eased slightly.

A *New Civil Engineer* supplement on the Middle East in 1977 reported: 'The commercial section of the British Embassy in Kuwait officially believes that the Kuwaitis want British contractors back.' But the representative in Kuwait of the UK firm Bovis, Tony West, was quoted as saying: 'This thing about wanting the Brits back has to be seen in the context of them wanting *everyone* back.'[1]

The legacy of the 1960s was that there was little continuity between earlier and later phases of projects.

Since the Gulf War many UK firms have been able to re-enter the market, winning reconstruction contracts which have led to further work.

Oil industry infrastructure
The British consultant Rendel Palmer & Tritton was present in Kuwait almost from the start of oil production. Oil exports began in June 1946 from two buoy berths located out to sea around 1.6 km from the beach,

served by two 300-mm diameter welded steel pipes. Another three were added later. But although rapid and cheap to construct such berths do not give a fast turn around of ships and are often unusable because of rough weather. It soon became clear that these buoy berths would not be capable of loading tankers as quickly as required.

Similarly, KOC's base for unloading stores and cargo was at Shuwaikh on the southern shore of Kuwait Bay, where there was only a small barge jetty to land cargo. Ships had to anchor 6.5 km offshore and cargo was brought in by lighters. Both facilities urgently needed to be replaced with something more substantial.

Rendel Palmer & Tritton was commissioned to advise. It evaluated various locations on the coast before deciding on Mina al-Ahmadi. The site had the advantage of being only 8 km from the oil company's head-quarters, workshops and residential areas which formed the new town of Ahmadi, whereas Shuwaikh was 38 km away.

The decision on the form of the new terminal was influenced by the urgency with which it was needed, since oil production was increasing rapidly. The timescale ruled out construction of a dredged harbour. Rendel Palmer & Tritton decided on a jetty berth since this would be less affected by adverse weather than the other options, dolphins or buoy moorings.

Design requirements included: capacity for a daily average export of 500,000 barrels of crude oil; capacity for monthly average imports of 19,000 t of cargo; accommodation for the simultaneous loading of six tankers and for the discharge of two general cargo vessels; a 12-m depth of water at all berths at low tide; and the fastest possible construction programme.

Because of the shallow slope of the seabed, the pier head had to be sited 1,280 m from the shore in order to reach water that was deep enough. A structure which is T-shaped in plan was adopted, with the approach from the shore forming the upright of the T, the oil jetty forming one leg of the top of the T and the cargo jetty the other. A 7.3-m wide road runs along the top of the jetty, supported on pile bents, or frames, at 4.6-m centres. Alongside, on extensions of alternate bents, runs the pipeway which then consisted of eight pipes of 600 mm diameter.

The 855-m long oil jetty has four tanker berths along the outer face and two on the inner. The longest is designed to accommodate 30,000-dwt tankers. The cargo jetty is 328 m long with a berth on each side. Steel H-piles were used to support the jetty, in the knowledge that corrosion was a potentially serious problem in the area which would have to be addressed.

Between June 1948 and October 1949 the 3,583 piles were driven using a single floating rig. Two additional piles were provided, for extraction after 10 and 20 years, to determine the condition of the steel beneath the seabed.

For the jetty structure, all welded construction was adopted with no bracing or other members below low water. Jetties elsewhere in the Gulf had been found to suffer rapidly from corrosion. At Mina al-Ahmadi, it was decided to protect all piling to within 3 m of the toe, and all structural members below pile cap level, with one coat of coal tar enamel applied hot before driving or erection. Even so, barnacles grew rapidly and stripped off the coating in some areas. However, it was concluded that, with the addition of cathodic protection, corrosion could be adequately controlled. Cathodic protection initially took the form of sacrificial magnesium anodes; later these were replaced with an impressed current system, which had cheaper running costs.

The oil jetty was commissioned on 23 November 1949 when three tankers berthed successfully in less than three hours. The cargo jetty was first used the following January.

After the Gulf War, Rendel Palmer & Tritton was commissioned to carry out a structural integrity survey of the jetty, including an inspection both above and below the water, and structural analysis. A programme of repairs and refurbishment was recommended.

Water projects

Throughout the 1950s, the Kuwaiti projects in which British firms were engaged were typical of a state with new found oil wealth. Sir William Halcrow & Partners undertook major harbour design. Ewbank & Partners and John Taylor & Sons worked together on power and water supply. While in the early 1960s Sir Frederick Snow & Partners were responsible for designing the international airport. The development of Kuwait was largely the development of Kuwait City, where the majority of the population lives. The city gradually acquired a spectacular water-front cityscape and a series of concentric ring roads as it spread out-wards from Kuwait Bay and the Gulf.

Development of water resources began shortly after World War II when the KOC appointed Ewbank & Partners as consultant to install a power station and desalination plant at Mina al-Ahmadi. Ewbank was also appointed by the amir to provide a power and desalination plant at Shuwaikh. While this was being built, a 150-mm steel pipeline was run across the desert from Mina al-Ahmadi to Kuwait City to augment the town's water supplies. Until then water had been brought in by sea in dhows from the Shatt al-Arab river at Basra, while methods of distributing the water had included goatskins on the backs of donkeys.

Integration of power and water became the norm in the Gulf, with multi-stage flash evaporation desalination plants using exhaust steam from the power station turbines. As desalination plants became more sophisticated, they were, by varying temperature and pressure, able to use the heat from the steam between 12 and 20 times over.

In 1951 Ewbank brought in John Taylor & Sons as subconsultant for water distribution and storage. The task was to provide a piped water-supply to four water-towers located at strategic points around the city, and thence to filling stations from where lorries took over distribution.

The first scheme provided for 4.5 million l/d of desalinated water and the same quantity of brackish water from a wellfield at Sulabiyah. Initial water-supply works were designed on the basis of a demand of 22 l/d per head. Demand has since increased beyond all expectations, as people have become accustomed to watering their gardens, and recent designs use a figure of 1,350 l/d per head. Restricting use is difficult as Muslims do not readily accept the idea of having to pay for what God has bestowed on them.

In the late 1970s Sir M. MacDonald & Partners took over responsibility for assisting the Kuwait Ministry of Electricity & Water's staff in designing phase II of the distribution system.

Between 1982 and 1988 John Taylor & Sons was responsible for design and construction supervision of treatment, storage, pumping and distribution works for water from the Az-Zour desalination plant. Most of Kuwait's desalination plants had been built near the border with Iraq. During the Iran–Iraq War, when Kuwait was providing funds to Iraq, there were fears that the desalination plants were vulnerable to attack from Iran, so the Az-Zour plant was built far to the south, near the Saudi Arabian border, as an insurance policy. Water was pumped 100 km north to Kuwait City in 1,200-mm diameter pipes.

The complex included enormous reservoirs and pumping stations to transfer and distribute 772 million l/d of water to Kuwait City. Altogether the project cost £250 million. The two pumping stations, each 100 m long, had control rooms at first-floor level. With the aid of the original plans, US forces in the Gulf War knocked out both, by smart bombing, the control rooms – apparently to the annoyance of the Kuwaitis, who would have liked one left intact.

John Taylor & Sons was also responsible at around the same time for planning and design on another Shuwaikh scheme, a water distribution complex for blending desalinated and groundwater to produce 455,000 m³/d of potable water. This has not yet been built.

Buildings

In the late 1970s and early 1980s, Ove Arup was responsible for several prestigious building projects. Two of these were for research related complexes, designed in conjunction with The Architects Collaborative of the US: a new headquarters and laboratory complex for the Kuwait Institute for Scientific Research (KISR) and a new headquarters for the Kuwait Foundation for the Advancement of Science (KFAS).

The KISR is the main government-funded body responsible for carrying out research into scientific fields deemed to be of particular importance to Kuwait. It has a multinational staff and has gained a high reputation for the quality of its work. Particular areas of interest are: oil and petrochemicals; desalination – it has built a reverse osmosis plant which produces 850,000 g/d of water but whose main function is as a research and development facility; and agriculture. Among achievements in this domain were the development of a simple-cell protein to be used as animal fodder, and of a plastic mulch which not only reduces evaporation and retards weed growth, but biodegrades at the end of the season providing nutrients for the plants.

The KISR's new 25,000-m² complex, for which master planning began in 1979, was built partly on reclaimed land to the west of the main port at Shuwaikh, close to the university. The buildings were arranged alongside atria. Laboratories were designed for research in engineering, earth and environmental sciences, food resources, petroleum and techno-economics. The complex had its own water treatment plant and an effluent treatment plant for neutralising laboratory wastes. Energy-saving measures included solar heating units.

The KFAS, founded in 1976, is complementary to KISR. It is funded by Kuwaiti shareholding companies, which contribute 5 per cent of net profits to the foundation; this is supplemented by donations from private individuals and families. It does not itself carry out research but sponsors others, not just in the Gulf but also in other Arab and Islamic states. It grants scholarships to individuals, provides funding to organisations for project work, and holds conferences and seminars at its headquarters building.

The brief for the KFAS headquarters, built between 1982 and 1986, required that it should express the fusion of art and science. It is a nine-storey, almost square, building with office floors arranged around a central atrium containing a 28 m Foucault pendulum. The roof over the atrium includes 10 large vertical drums through which the amount of natural daylight is controlled by rotating polarising filters.

Buildings such as this, and the Salhia complex, which comprises a commercial centre and the Meridien hotel, demonstrated that high standards of construction could be achieved in the Middle East. According to *The Arup Journal* in 1979, the Salhia complex showed that: 'In terms of building structure, many of the up-to-date techniques of Western industry are being used successfully in the Middle East. Much has been written and heard of poor quality construction in that part of the world, and it certainly exists. However it is equally true that, with proper attention, work of the best quality can readily be achieved.[2]

The commercial centre at the Salhia complex consisted of a long block containing five storeys of offices over three of shops and two of parking. The Meridien hotel has a tower containing 13 floors of bedrooms with a nightclub and swimming pool on top, with restaurants, conference and meeting rooms below.

Both buildings were founded on rafts on Kuwait's typical subsoil of compact, lightly cemented sand. The water-table lay 1.5 m above the bottom of the two-storey basements, and wellpoint dewatering was used. The site was closely bounded by a road and a seven-storey building, which created the need for reasonably complex temporary propping details to prevent these structures being undermined by the excavations.

The superstructure of the commercial centre is divided into five sections, with a grand entrance in the central section. Twelve lift/stair cores which stabilise the structure were slipformed, though slipforming had to stop during the hot summer months from May to September. One core had to be abandoned when the concrete dried so quickly in the heat that the moving shutter dragged out large cavities and cracks. This core had to be cut down and restarted after the summer.

Floors were prestressed, precast concrete T-section floor units, with a structural topping, mostly spanning 18 m. The units were locally made in a new factory set up for the project.

The hotel tower has structural crosswalls to the bedrooms at 4-m centres. Below the rooms a thick transfer slab takes the loads of these walls and allows the structure to be opened up to an internal grid of 8 m x 8.8 m to provide space for the public rooms. This was cast in two layers at the request of the contractor, Ahmadiah Contracting; the first, of 500 mm was poured onto normal falsework and allowed to strengthen until it could support the second pour of 1-m thickness. French Outinord tunnel formwork was used to construct the upper floors. This folds outwards, supporting the slab at midspan. The contractor overcame initial reservations and adopted the system enthusiastically.

Pumped concrete was used extensively. The construction programme called for a floor to be completed every 12 days: the contractor achieved a nine-day average. The structure was completed midway through 1979, about 21 months after work started. Structural concrete was of a standard as high as could be achieved anywhere, Ove Arup engineers believed.

Agriculture and irrigation

Agriculture is not a significant part of Kuwait's economy, although, as mentioned above, big strides towards self-sufficiency in some areas have been taken. Only 10 per cent of the land is cultivable, and by no means all of this is under cultivation. Much land is suitable only for pasture, and lack of water is as much a problem as unsuitable soils.

However, between 1978 and 1985 John Taylor & Sons was engaged on a project to address the water part of this equation, by planning and designing works to enable reuse of sewage effluent in agriculture and forestry.

The realisation that sewage effluent was a potentially valuable resource in the arid Middle East coincided with the development made possible by oil wealth. John Cowan and Paul Johnson of John Taylor & Sons, in a paper to the 1984 ICE conference on the subject, said:

> The growth of new urban communities with their demands for modern services of every type in such a hostile climatic environment has created an awareness of the need to deploy every resource to its ultimate extent. This has been particularly illustrated in the context of water resources where the installation of piped water supplies and sewerage systems had created a supplementary resource – waste water, offering prospects of reuse if treated sufficiently.[3]

Effluent standards to enable reuse are set out by the WHO but many Middle Eastern states chose to aim for higher standards than these. The prime objective of treatment is to prevent the risk of a public health threat. Health hazards can be transmitted via water by microbiological or chemical contamination.

Microbiological contamination includes bacteria, viruses, protozoa, and helminths. Bacteria are responsible for diseases such as typhoid, cholera, bacillary dysentery and salmonella poisoning. They are much reduced by conventional wastewater treatment but chlorination is usually required to eliminate them. Information on transmission of viral disease through wastewater is limited but diseases which could be transmitted include infectious hepatitis, polio, and enteric diseases. Viral content of wastewater is generally lower than that of bacteria but viruses are more resistant to treatment processes. It is believed that viruses cannot survive more than seven days in effluent, however.

Protozoa such as amoebae can cause amoebic dysentery. They can be removed from effluent by tertiary filtration. Helminths are worms or flukes causing debilitating diseases and are often parasitic. Transmission is by eggs which can survive for long periods. They can be deposited on soil or crops through irrigation and picked up by physical contact. Most eggs can be removed from effluent by a combination of sediment and tertiary treatment.

The main means by which pathogenic organisms arising from effluent are transferred to humans is via irrigated fruit and vegetables taken from field to market to the home. The greatest risk is from food eaten raw since most pathogens are killed by cooking. Chemical contamination is not a great problem in the Middle East because there is relatively little industry there.

Recommended irrigation methods take account of the routes by which public health could be affected. Thus, for crops which are normally cooked before eating, sprinkler irrigation can be used. For crops which are normally eaten raw, a more sophisticated non-contact system such as drip-feed is to be preferred.

Effluent reuse began in Kuwait on a limited scale in the mid-1970s. Towards the end of the decade, John Taylor & Sons was commissioned to produce a master plan for effluent reuse in agriculture, the aim being to make Kuwait self-sufficient in milk, potatoes, onions and garlic by the year 2010. The scheme, which began to be commissioned during the 1980s, involved the transfer of treated effluent from three sewage treatment works to a central distribution and administration centre. Here it was directed to an existing farm and to a newly created agricultural area, with third area to be developed later. Tertiary treatment consisted of rapid gravity sand filtration and chlorination. The irrigation distribution network was designed to handle 450,000 m³/d of treated effluent by 2010.

The established farm covered 860 ha and to this a new area of 870 ha was to be added, with a further 1,400 ha in the future. The main crops were alfalfa for the dairy industry, onions and garlic, and, on an experimental basis, aubergines and peppers. In addition, it was planned to cover an area of 12,000 ha with forest. The scheme was set to be, by 2010, one of the biggest effluent reuse schemes in the Middle East.

Roads

As Kuwait City expanded during the late 1970s and 1980s, a vast number of roads was built. It now boasts seven 'ring' roads at progressively greater distances from the centre, though because the city has radiated out from a point on the coast, only the innermost is a complete ring – the rest are only half-rings. Kuwait's ministry of public works enlisted the help of The US Federal Highway Administration to develop its network. US consultant De Leuw Cather produced the master plan for the ring-road system.

British consultants Freeman Fox & Partners and Mouchel were active in designing Kuwaiti roads during this period. A typical project was Mouchel's Riyadh Street Expressway which is a radial route running south from the centre and crossing the Second, Third and Fourth rings. It is a 5-km dual three-lane motorway with central concrete 'New Jersey' crash barriers. It crosses the Second and Third rings on 280-m seven-span post-tensioned viaducts. At the Fourth ring there is a complex intersection with Riyadh Street in a cutting, the Fourth ring crossing it on a flyover, and a gyratory system connecting the two at ground level. The flyover had to be kept in continuous use during construction to prevent traffic disruption, so it was underpinned with in-situ concrete bored piles which also form the retaining walls of the cutting.

In 1979 Mouchel designed 8 km of the 11-km long first ring. The dual three-lane motorway includes seven grade-separated interchanges, a 1.5 km viaduct and 850 m of dual two-lane tunnel beneath the city's central business district. In addition 30 km of local roads were upgraded; junctions; between this network and the motorway slip-roads are controlled by 50 sets of traffic-lights. Of the local roads, 3 km were built on reclaimed land.

Much of the motorway was built in cutting, which, because of the high water-table, meant that special precautions were needed to prevent uplift. The tunnel under the central business district was designed to have bored piled walls, with a reinforced concrete roof supporting access roads which cross the tunnel at ground level. Construction was divided into seven contracts, of which the first began in 1983. Incorporated into the roadworks were 15 km of the main stormwater system and five new sea outfalls.

Overall completion was originally scheduled for 1990 but construction of only four contracts had been started by then. Work was still under way on parts of the ring road at the time of the Iraqi invasion. Serious flooding of several underpasses happened after dewatering stopped when work was suspended. After the war Mouchel was appointed, with the local firm Gulf Consult, to produce a damage assessment report and to prepare proposals for remedial and completion works.

Freeman Fox & Partners was commissioned in 1978 to design and supervise construction of two motorways, one being part of the southern route from Kuwait City to Saudi Arabia and the other part of a western route to Iraq, Jordan and Europe. The two projects included a total length of 55 km of which 9 km was dual four-lane, the rest being dual three-lane. There were 12 grade-separated interchanges involving the construction of two 11-span, 450-m long viaducts and 15 major highway bridges. These were all post tensioned in-situ concrete box girder structures.

Freeman Fox & Partners was also responsible for improving the 36-km radial Jahra–Ghazali motorway, to a dual three- and four-lane expressway with grade separated interchanges. Because of severe restrictions imposed by existing development, and to keep the heavy traffic flowing, 10 km of the motorway was built on viaduct over the existing arterial road. The viaduct was a precast prestressed segmental box girder.

By the mid-1980s there were half-a-million vehicles on Kuwait's roads, and, unsurprisingly given the sudden transformation into a motorised society, a considerable road safety problem had arisen. The government launched an extensive education campaign to combat this.

Kuwait today

Following the Gulf War rebuilding work was let by the US Army Corps of Engineers and the amount won by British firms was initially disappointing. Nonetheless, there were some successes and as a result British contractors are now probably more active in the state than at any time since its independence; the initial war repair work has led to other contracts both in Kuwait and elsewhere for the successful companies.

British contractors, including Taylor Woodrow International, Amec and Wimpey formed a joint venture, the Kuwait British Fire Group, and won contracts worth US$77 million for general fire-fighting work and for the repair and refurbishment of the war-damaged gathering centre GC-23. The joint venture was assisted by BP, Royal Ordnance and other specialists. Taylor Woodrow and Wimpey, working as Kadcol – Kuwaiti-Anglo Defence Company – have since won further contracts, including a US$36 million contract for repair and refurbishment of Kuwait's 125-ha naval base.

Morrison Shand won contracts for restoring water and sewerage systems to working order after the war, and this has led to a £4.9 million reservoir construction contract at Umm Ghudair, in association with Kuwaiti contractor Boodai. The firm has also gone on to win the contract to refurbish the Manama–Sitra causeway in Bahrain (see chapter 4) and has registered a company in Dubai.

Most of the damage done to the country's infrastructure during the war has now been repaired. All three oil refineries are back in operation to some extent, at a cost of KD120 million, though throughput by early 1994 was still only 645,000 b/d compared with 776,000 b/d before the war. The absence of Palestinian and other expatriate workers has been felt.

It is not known whether the uncontrolled release of oil, in the inferno started by the retreating Iraqis, will have caused any permanent damage to the Burgan field. A peak of 6 million b/d was escaping from the field, more than the daily production of the UK sector of the North Sea, and a total of a billion barrels was lost. Some of this remains in 200 lakes formed by oil which escaped from wells that failed to catch fire; these are being drained and the oil will be converted to bitumen. How to deal with up to 20 million m^3 of contaminated soils is still under discussion.

A more pressing problem is cash flow. Kuwait had to liquidate a considerable part of its foreign investment funds to pay for the war and for rebuilding – though its assets remain considerable – and depressed oil prices mean that oil revenue is low. Government ministers are calling for 'austerity' and for cuts in public spending, including the extensive welfare state provision. This coincided with the election, in 1992, of the toughest and most independent-minded national assembly ever, which set about scrutinising ministerial decisions with unprecedented vigour making the prospect of a paralysing stand-off between the two seem likely.

c h a p t e r   s e v e n

QATAR

The state of Qatar is a small, mainly desert peninsula extending 180 km into the Gulf. It has a population of only about 350,000. A former British protectorate which became independent in 1971 at the same time as the United Arab Emirates, it was involved in the initial talks aimed at setting up the planned federation of nine Gulf states. However, like Bahrain, it later withdrew from the negotiations, its size being just enough for it to be viable as an independent state.

Qatar has been an oil producer since just after World War II. Its oil resources are relatively modest, but the state also boasts one of the world's largest natural gas fields, whose development is likely to assure its prosperity for a considerable time to come.

There is archeological evidence of settlements in Qatar as long ago as 4,000 BC, but its history is not well documented. The whole of the Arabian Gulf coast was an important trading area around the eighth century AD; the inhabitants of Qatar also made a living from pearl-fishing.

Its inhospitable climate and terrain made Qatar unattractive to most of the European empires, who ignored it, and it was dominated by Bahrain until it became part of the Ottoman empire in 1872. As Ottoman rule declined, Britain signed protection treaties with the dominant al-Thani family in 1916 and 1934.

An oil concession was granted to the Anglo-Persian Oil company in 1935 and was later transferred to Petroleum Development (Qatar) Ltd. Oil in commercial quantities was discovered just before the outbreak of World War II. Exploitation was held up until 1949 when production began from the Dukhan field.

Qatar was initially enthusiastic about negotiations to form a federation of nine Gulf states (comprising the seven emirates which are now the United Arab Emirates, with Qatar and Bahrain) following Britain's decision to pull out of the region in 1968. But following disagreements Qatar opted for independence as a separate state.

As elsewhere in the Gulf, oil revenue allowed development of the state's infrastructure which is now largely complete. The economy has grown rapidly since 1960. Oil reserves are, however, thought to be sufficient for only 25 to 30 years' production at current rates. The government has, since the early 1970s, aimed to diversify the economy by setting up industries such as the fertiliser plant owned by Qatar Fertiliser Company (QAFCO), the steelworks owned by Qatar Steel Company (QASCO) and the petrochemicals plant owned by Qatar Petrochemical Company (QAPCO).

But Qatar's greatest asset is the huge offshore gas reservoir which lies off the north-east coast and contains recoverable reserves estimated at 4 per cent of the world total. Development of this field, discovered in 1971, is the state's biggest project at present.

Construction projects have not been as extensive in Qatar as elsewhere in the Gulf. *New Civil Engineer* commented in a supplement published in 1977: 'The Qataris are prudent people, resembling in character their Saudi neighbours more than their Gulf compatriots. They display few extravagances and do not like wasting money.'[1] If anything, this prudence surpassed that of their neighbours. Port congestion in the late 1970s was as great a problem in Qatar as elsewhere, and *New Civil Engineer* reported that the Qataris were making strong efforts to clear the congestion by enlarging existing facilities. But it added: 'Reflecting sensible Qatari caution, plans to build a vast new port north of Doha, which the government suspected would give the country surplus berths, are reported to have been quietly shelved.'

Water supply

Qatar has no rivers and its annual average rainfall is only 50 mm. Even in the early 1950s, when oil exports had just begun and the population was estimated at only 20,000, water supply was a problem. The shortage was particularly acute in the capital, Doha, where, at the time, about half the population lived. (The proportion has since risen to three-quarters.)

When oil revenues started to flow, the amir decided to set aside 25 per cent of the total for economic development to improve living standards. Priority was given to roads, schools, hospitals, electricity supply, and the provision of an adequate water supply for Doha.[2]

Doha was at that time supplied with water from wells several miles outside the capital which were owned by private operators who sold to the inhabitants. This was a factor in limiting consumption to only 11 l/d per head. In 1952 J. D. and D. M. Watson was appointed to advise on water supply. It aimed to provide an absolute minimum of 23 l/d per head for 12,000 people by desalination of sea water, and hoped to double this by developing groundwater sources. The eventual aim was to provide 90 l/d per head.

It was also planned, later, to duplicate the system if, as then seemed unlikely, the population expanded beyond the initial design figure. Oil exploration had revealed no large aquifers of good quality water, such as those which later provided the basis for the Great Man-Made River project in Libya, and though groundwater could be found by sinking wells to around sea level almost anywhere in the peninsula, its quality was unpredictable.

However, geological information suggested that a limited area around Doha could yield sufficient water and this proved to be the case. The advantage of having two sources, each of which was capable of producing a reasonable minimum quantity of water, was obvious; in addition, by blending distillate from the desalination plant with the slightly brackish groundwater, a blend more palatable than that of either source taken separately could be obtained.

A suitable site for a first wellfield was found 16 km from Doha and exploitation began as a matter of urgency, since a distillation plant would take some time to design and build. Four wellfields up to 3 sq km each in area were developed. Provision was made for wells to be 'rested' in rotation because it was found that water quality deteriorated rapidly if overpumping occurred.

Distillation plants were at that time relatively uncommon. The winning tender was for a three-unit, triple-effect evaporator plant, capable of producing 450 m^3 of distillate a day. The plant used oil-fired boilers which could be converted to natural gas, in anticipation of cheap supplies of gas becoming available. The sea water is highly saline and contains high concentrations of calcium and magnesium salts. It is therefore pretreated with an organic dispersant before distillation, to limit deposition of scale. The plant was fully operational by January 1955.

Water was distributed to 545-m^3 tanks from where Doha's population collected it in cans from standpipes. A nominal charge was made to discourage waste.

The population expanded much more rapidly than envisaged and when it reached 18,000 in 1954, phase B duplication had to be put into effect. The second, similar, distillation plant was brought into service in 1956 and groundwater abstractions were also increased.

Watson's involvement ended in the late 1950s. In 1962 Pencol took over responsibility for water supply, while a joint venture of Pencol and John Taylor & Sons was responsible for wastewater.

As the population continued to increase, more, larger desalination plants were built to meet demand. Between 1967 and 1978 a new central distillation plant was built on the coast at Ras Abu Aboud. The total capacity of the multi-stage flash plant eventually reached 45,500 m³/d.

The first two units, forming phase A, provided 4,550 m³/d and were completed in 1969 at a cost of £1 million. They operate on steam from an adjoining power station. Phase B, built between 1971 and 1973 and costing £4.8 million, added two more units with double the capacity of phase A, originally using polyphosphate or acid dosing for scale control. In 1974, non-acid high-temperature scale-control additives became available and the phase B units were among the first plants in the world to be converted to their use. Output was increased to 11,360 m³/d at the same time.

The third phase saw the addition of two further units with the same capacity as phase B; this was completed in 1978 at a cost of £17.4 million.

Over roughly the same period, from 1967–1974, further developments to increase output from wellfields 40–50 km north-west of Doha took place. Pencol designed pipelines and pumping stations to convey this water to the city. Phase I comprised four pumping stations and pipelines to Doha. Later, a new wellfield was developed at Musruah, and a balancing tank and booster pumping station were added. The supply reached 9,090 m³/d in 1978.

Another combined sea water desalination plant and power station was built in phases on the coast at Ras Abu Fontas, south of Doha, between 1974 and 1983. Pencol was responsible for design and supervision of distillate storage reservoirs, pumping plant and pipelines. Phase I comprised an 18,200-m³/d reinforced concrete storage reservoir with a pumping station and a 17-km long, 1,200-mm diameter pipeline. A second reservoir of the same capacity was added in 1980; later a further phase, including four 36,400-m³ reservoirs and an additional 19-km pipeline, was added as the total output of the plant increased to 273,000 m³/d.

Water is also pumped to Umm Said industrial zone from Ras Abu Fontas, for which an additional 34-km, 60-mm pipeline was built. Storage reservoirs, pumping stations, a water-tower and chlorination facilities were also provided at Umm Said. Pencol was responsible for distribution schemes for several other towns throughout Qatar.

Wastewater

From the 1960s onwards, Qatar's extremely fast rate of population growth (from 70,000 in 1967 to 250,000 in 1984), especially in the capital, meant continual expansion of Doha's wastewater system. Pencol and John Taylor & Sons were consultants from 1966. A sewage works providing primary and secondary treatment, to serve a population of 67,000, came into operation in 1968–1971. It was later extended to serve a population of 90,000.

By 1977 it was apparent that further expansion was needed, and at the same time the consultants were asked to provide tertiary treatment facilities to bring effluent quality to a standard suitable for use in agriculture and municipal beautification. The consultant recommended a completely new works at Naijah, south of Doha, designed for a population of 100,000. This would use the activated-sludge process rather than filtration, as was used in the original treatment works. The plant, which was completed in 1983, produced an effluent of 20-mg/l biochemical oxygen demand (BOD) and 30-mg/l suspended solids.

It was later expanded to serve double the original design population. Tertiary treatment based on rapid gravity sand filtration, with sophisticated telemetry and monitoring facilities, was also added, bringing effluent to 10 mg/l for both BOD and suspended solids.

As the terrain is extremely flat, the sewage has to be pumped to the treatment works. Over 70 pumping stations have been built, varying from small submersible stations to main forwarding stations with up to six multi-speed pumps, each with a capacity of 300 l/s. Preferred pipe materials are vitrified clay for gravity sewers and ductile cast-iron for rising mains.

Effluent is pumped to several elevated storage towers around Doha, connected via a ring main, and to a government farm 45 km away. Within Doha it is used on the verges of major roads and for landscaping. Final distribution is mainly by road tanker but there are plans for additional tanks and distribution mains to areas where the irrigation water is needed.

Meanwhile, further expansion and development, particularly in the north and north-west of Doha, led to the construction of another works 20 km west of the capital to serve another 100,000 people and capable of subsequent expansion to serve 400,000. The initial works was completed in 1991 and the construction of the first extension, to serve a population of 150,000, began in 1993. The aim of the sewerage division of Qatar's ministry of public works is to have most of the population connected to a modern sewerage system by the mid-1990s.

Ports

Qatar's first venture in economic diversification came with the development in the late 1960s of the fertiliser plant which produces ammonia and urea for export from natural gas.

Sir Alexander Gibb & Partners was responsible in 1968 for a feasibility study for the fertiliser plant. This included market forecasts, economic analysis, site selection, and estimates of cost and cash flow. The chosen site was at Umm Said, where it was also decided to build a flour mill.

Following completion of the study, QAFCO (Qatar Fertiliser Company) was set up and Gibb was commissioned to carry out design, specifications, construction supervision and commissioning for the first stage of the project, built between 1968 and 1974. The plant was to have a capacity of 1,000 t/d of urea and 900 t/d of liquid ammonia. Included in the project was a 50-MW power station and desalination plant, a cooling system using sea water, a workshop, laboratory and offices.

In 1975, Gibb was appointed subconsultant to Norsk Hydro of Norway for the design and supervision of civil and building works for Stage II of the project. This stage, which came on stream in June 1979, doubled the plant's capacity.

At the same time as Stage I of the fertiliser plant got under way, Gibb was appointed by the Qatar government to design and supervise construction of a jetty to serve both the fertiliser plant and the flour mill.

The jetty extends into water 12 m deep with a berth on each side, one equipped with pipelines to export liquid ammonia and conveyors and loading arms for bagged or bulk urea. The other is equipped to unload bulk grain. The jetty also serves as a water intake for the fertiliser plant's cooling system.

In 1975 Gibb went on to produce a master plan for development of a new industrial port at Umm Said. The existing jetty was incorporated into the port facilities as berths 20 and 21. The site is a long, gently sloping, sandy beach protected by an offshore reef.

Construction began in 1975 with dredging to provide an approach channel and turning-circle of the required water depth, and to reclaim 650 ha for the industrial development. A total of 23.3 million m^3 of material was dredged. A start was also made on the tied-back sheet pile walls for the small craft harbour and berths 9 and 10 for general cargo.

The industrial area was to accommodate a steelworks and petro-chemicals plant. Berths 1 to 3 are equipped to unload iron ore in either fines or pellets at the rate of 1,000 t/h. Berths 4 to 6 serve the new steelworks and have four 12-t cranes, two for grabbing and two for handling steel products. Berths 18 and 19 are for handling petrochemicals' products.

Cellular sheet pile cofferdam construction with 20-m diameter cells was adopted for most of the berths; 1 to 6 were built between 1976 and 1977, followed by 18 and 19. Berths 7, 8 and 11 to 17 were to be constructed as a later phase of the development.

Buildings

In 1973 the first students entered the new University of Qatar. The intake of 150 was housed in temporary buildings. By the mid-1980s the student population had grown to 4,700 and was expected to continue rising. Permanent academic buildings, comprising phase 1A of the development plan for the university and totalling 73,000 m^2 in area, were inaugurated in 1984, five years after construction began. The contract value was £125 million and the main contractor was Fujita of Japan. The buildings were designed by Egyptian architect Dr Kamal El-Kafrawi, with Ove Arup & Partners as structural and services engineer.

Planning of the buildings was based on a twin grid of 8.4-m wide octagons connected by 3.5m squares. The octagon is a shape traditionally used in Islamic architecture and design. They are low-rise, modular and constructed in concrete. The modular form allowed for extensive pre-casting both for structural elements and the high-quality white precast cladding panels. Precasting allowed a high standard of quality control and speeded construction. The buildings use traditional Arab wind towers over classrooms to allow indirect natural light and ventilation air into the building. As a result air-conditioning is only needed for the hottest months of the year.

The primary structure consists of precast concrete loadbearing walls with waffle or trough in-situ concrete floor slabs. The roof is also generally formed from precast slabs, with two exceptions. Over laboratories, which are formed by combining two octagons, Vierendeel precast concrete roof trusses span the double octagon. Courtyards are formed by combining four octagons, and these are spanned by an in-situ folding-plate roof.

Services are distributed from a central services unit with an underground walk-through service duct connecting it to the other buildings.

Phase 1A included classrooms, laboratories, lecture theatres, a library and exhibition halls, administration offices, a canteen and the central services unit. A dual-carriageway ring road allows traffic to circulate round the campus. Phase 1A, which followed two years after the first, included sports facilities and other buildings.

One of the more unusual projects on which British companies have worked in the Gulf was the rescue operation for Qatar's stadium for the Asian Games. In November 1975, just four months before the games were due to open, an eleventh-hour bid was launched to finish the stadium. The Qatari department of public works called in Ove Arup, contractor Turriff-Taylor, and architect Pack-Roberts. The main structure, a 40,000-seat stadium, was apparently suffering major quality control problems. 'Quality of workmanship had fallen down just about totally, due to a complete lack of co-ordination and [qualified] site supervision,' said a source quoted in *New Civil Engineer* at the time.[3]

Baghdad based Iraq Consult, which had been responsible for the design but not allowed on site to supervise, had been relieved of its duties completely. The local contractor Al Atiyah remained on site working in conjunction with Turriff-Taylor.

The stadium was a reinforced concrete structure with an in-situ frame and precast cross members supporting the seats. The Ove Arup–Turriff-Taylor team discovered that quality control on the concrete, vital in the heat of the Gulf, had been almost non-existent; high yield and mild steel reinforcement had been arbitrarily interchanged; bars had been positioned out of tolerance; and control over concrete placing had been lacking.

Ove Arup assigned 19 engineers to the site, backed up by another 12 in London, entrusted with the task of putting matters right, while Turriff brought in 40 supervisory craftsmen and considerable quantities of equipment. A supreme effort by the site staff, who were reported at one stage to be putting in double their usual hours, made the operation a success and the stadium was ready for the March opening.

Qatar today

Qatar has enough oil to last only another 25 to 30 years. For this reason it was decided to go ahead with the massive programme to exploit and export natural gas from the field lying off its north shore and known as the North Field. The field's proven reserves are 6,400 billion m^3, ten times as much as the UK's reserves in the North Sea and enough to last perhaps 200 years, according to experts.

The infrastructure needed to export natural gas is, however, more complex and expensive than that required for oil. Qatar is investing huge sums to provide that infrastructure at Ras Laffan, where the world's largest port construction project is currently under way.

This is due to be completed in 1995, at a cost of US$750 million. An Italian consortium of Condotte, Dragomar and Grandi Lavori Fincosit began work on a five-year dredging contract for the harbour in 1991. French contractor Bouygues, with its subsidiary SB Technigaz, won the US$120 million contract for three 85,000-m^3 gas storage tanks. The first project to export liquefied natural gas (LNG) is expected to come on stream in 1997, when Qatar Gas will start exporting 4 million t of LNG annually to Japan, from a US$1,600 million liquefaction plant. The following year the larger Ras Laffan Gas project is due to begin exports of 9 million t of LNG a year to South Korea and Taiwan. Following a bloodless coup in June 1995 in which Sheikh Hamad bin Khalifah al-Thani deposed his father Sheikh Khalifah bin Hamad al-Thani, development of the North Field was expected to be pursued more aggresively.

UNITED ARAB
EMIRATES

The seven emirates of Abu Dhabi, Dubai, Sharjah, Ras al-Khaimah, Umm al-Qaiwain, Fujairah and Ajman are perhaps the most unlikely participants in the economic transformation of the Middle East. The largely desert emirates have a coastline which is long compared with their overall size. Part of the coastline is within the Gulf and part on the Gulf of Oman, providing access to the Indian Ocean. Both coasts shelve gently to the sea. Inland are uncultivable salt marshes giving way to desert. For these reasons settlements are concentrated on the coast, and fishing, pearling and trade became the chief economic activities. There were two main tribes there, the Bani Yas who dominated Abu Dhabi and Dubai, and the Qawasim, who inhabited Ras al-Khaimah and Sharjah.

The discovery of oil in Abu Dhabi, Dubai and Sharjah opened the way for their subsistence economies (dependent on fishing and pearling) to become industrialised. But Abu Dhabi, the largest of the emirates, had only a tiny population when oil was discovered, so that it would have been difficult for it to develop a viable economy alone.

This prompted the formation of the United Arab Emirates when Britain withdrew from the Gulf in 1971, a federation aimed at producing cooperation where before there had been rivalry, antagonism, and border disputes.

But the emirates had attracted attention long before there was any hint of oil wealth. The strategic importance conferred by their geographic location was recognised in the 1800s when Britain entered into a series of maritime truces with their rulers.

The truces which gave the emirates the name 'Trucial States' were instigated by Britain to stamp out piracy on the shipping routes from Europe to India. The first was between Britain and the Qawasim in 1806. Under this accord the Qawasim agreed to respect the East India Company's flag in return for being allowed to use ports on the Indian coast. The first formal treaty with the states followed in 1820. Fifteen years later Sharjah, Dubai, Ajman and Abu Dhabi agreed a truce under which Britain took responsiblity for retaliation against any act of aggression against them. This agreement was renewed at intervals until 1853 when a perpetual maritime truce was proclaimed.

In 1892 Britain went further and signed 'exclusive' accords with the rulers under which they agreed not to enter into any agreement with, or sell any part of their territory to, any other power. Despite these agreements, Britain became worried in the 1930s that one of the rulers would grant a concession for oil prospecting to an exclusively US concern, as had happened in Saudi Arabia. A company called Petroleum Concessions, with English, Dutch, French and US partners, was therefore formed in 1935 to negotiate with the rulers. Dubai, Sharjah, and Ras al-Khaimah granted concessions, under pressure, in 1938. Sheikh Shakhbut of Abu Dhabi held out for better terms until 1939, convinced that there was oil in his territory because it had been seen bubbling to the surface around some of Abu Dhabi's islands. He was also anxious about the effect sudden wealth might have on his sheikhdom.

Exploration and development were held up by World War II and then concentrated on Qatar and Kuwait where oil in commercial quantities had already been discovered. When oil development resumed in the emirates it was disrupted by internal feuding. Britain took a more active role, providing a peacekeeping force, the Trucial Oman Levies. It also set up the Trucial States Council to encourage the rulers to cooperate, and provided development money.

Border disputes were common, because before the search for oil no serious attempt to define any boundaries had been made. Ajman, for example, consists of four separate territories, one administered jointly with Oman. Fujairah is also split into four parts, one jointly governed with Sharjah. Abu Dhabi and Dubai were at war in 1945–8. There were also disputes between the emirates and their larger neighbours. For instance, Saudi Arabia for many years laid claim to land around the Buraimi oasis, which would, if recognised, have deprived Abu Dhabi of some of its most important onshore oilfields.

Oil was discovered first in Abu Dhabi in 1958 (from where it was exported from 1962 onwards), and next in Dubai in 1966 (with exports beginning three years later), while in Sharjah production began only in 1974. Sharjah's oil revenue is shared with Umm al-Qaiwain, which receives 30 per cent, and Ajman, which receives 5 per cent. Abu Dhabi and Dubai also have large quantities of natural gas. Oil has more recently been discovered in Ras al-Khaimah.

The formation of the United Arab Emirates was prompted by Britain's decision in 1968 to close all its military bases east of Suez in order to reduce defence expenditure. Initially the federation was to have included Qatar and Bahrain. Qatar suggested that the five smaller emirates should first merge, in order to even out the disparity in size between the potential members, but this suggestion proved unacceptable. In the end Qatar and Bahrain opted for separate independence and, just before British withdrawal in July 1971, six emirates agreed to form a federation. Ras al-Khaimah joined a few months later, in February 1972. Just before the British withdrawal Iran occupied the Tumb islands whose ownership it disputed with Ras al-Khaimah. Because Britain took no defensive action, which under the perpetual truce it was in principle bound to do, Iraq cut diplomatic relations with Britain in protest, while Libya nationalised British Petroleum's interests there.

At the time of federation, Abu Dhabi was the only emirate to have a well-developed oil industry, and this increased its dominance. As the largest and richest emirate it became the capital of the federation. However, while Abu Dhabi has generally followed OPEC rulings and cut its oil output when required (since it has only a small population it can afford to reduce production to maintain prices), during the 1970s and 1980s Dubai tended in contrast to go its own way, so that oil production in Dubai rose closer to that of Abu Dhabi, and its industrialisation plans were expanded while Abu Dhabi's were scaled down.

In the early days of oil production, construction work followed the familiar pattern of providing basic infrastructure such as ports, roads (which were almost non-existent), water and power supplies. Later attempts to diversify the emirates' economies were hindered by lack of coordination and duplication of projects. Each emirate wanted its own deep-water port, international airport and so on; in 1986 the emirates' combined cement-making capacity was three times the domestic markets requirement. A five-year plan adopted in 1981 attempted to end this duplication.

Dubai has, perhaps, the most diversified economy in the emirates, built on its historic status as a trading port. It has an aluminium smelter, and the huge Dubai dry dock – big enough to hold the largest tankers ever built – and has set up a free trade zone at Jebel Ali. Its 39-storey international trade and exhibition centre, a Canary Wharf-style building costing £175 million at today's prices, designed by architect John R. Harris & Partners and built by Bernard Sunley, epitomises its aspirations to be a world trade thoroughfare.

Abu Dhabi and Dubai are also remarkable for the extent to which they have defied their arid climate, using widespread irrigation to beautify urban areas with extensive green landscaping. In Dubai this has extended to the construction of a golf course and horse-racing tracks. Dubai and Abu Dhabi also have ice-skating rinks.

Traditionally Dubai has been renowned, for the maritime trade centred on the creek and for buildings that incorporate wind towers, and which use wind and natural convection currents to lower ambient temperatures inside the buildings.

A b u   D h a b i

Ports

As elsewhere in the Gulf, a modern deepwater port was one of the first priorities for Abu Dhabi when the start of oil production generated increased economic activity.

In 1968 Sir Alexander Gibb & Partners was appointed consultant to the Abu Dhabi Ministry of Public Works and Development, later the Public Works Department with a brief to produce a master plan for development of Mina Zayed. The company has been working on this port ever since. 'When we started in summer 1968 there were just a few ramshackle jetties,' says Gibb's technical director and head of maritime engineering Peter Hunter.

The main constraint was the gently shelving shoreline. 'Ships had to anchor several miles offshore and unload into lighters. It was very slow and inefficient and expensive.' Gibb was given two priorities: first, to build a lighterage wharf as quickly as possible to bring about an immediate improvement in unloading times; and second to review an earlier master plan produced by the Canadian firm Cansult.

Shallow water at this point in the Gulf extends 6 km offshore. Cansult had suggested building a long causeway to two deepwater berths. Aware of the constraints associated with building berths at the end of a causeway, which could itself become a bottleneck, Gibb considered the possibility of building a deepwater port closer inshore. This would require extensive dredging to create a 7-km long navigation channel and an anchorage and manœuvring space in the harbour, but would allow more scope for efficient and economical future expansion.

The initial plan included a breakwater and nine berths, with scope for the later addition of 10 more, and another 10 after that. 'It was important that a lot of the projects had the scope for development that they needed, so that rather than build a short-term solution in 1968, something was provided that could be developed into what they really needed 10 or 20 years later. Against that there's the criticism that white elephants could be created,' says Peter Hunter. Often clients wanted to build something far bigger than they needed in the foreseeable future. 'Sometimes they're right and their biggest ambitions seem to come true, though at the time they seem unreasonable. We've been perhaps more cautious here than other places. When Jebel Ali [port] was built that had 70 berths and I think still has spare capacity.

A crash programme was immediately put into effect to build the 460-m long lighterage wharf which was designed and built in eight months and brought into service in early 1969.

The first deepwater facilities became operational in 1972 and included 1,164 m of wharves with six berths and the 3,000-m west breakwater. Dredged material from the 9.5-m deep, 7-km long navigation channel was used to reclaim land on which were built transit and storage sheds, and administrative buildings. A total of 7 million m^3 of loose and cemented sand was removed.

An update of the master plan followed in 1974. This led to construction of berths 10 to 19, four of which were for harbour craft, five for general cargo and one for grain handling, serving a flour mill built behind the berth.

Since 1976 the two original berths on the lighterage wharves have been converted to deepwater wharves and have become part of the port's container terminal. Also three additional deep water wharves have been constructed. The harbour and approach channel were deepened to 11.5 m in 1979, providing reclamation material for a further 10 berths and the dhow harbour, which was completed in 1981. Since then a fishing harbour has been added which was completed in 1992. There was no provision for bulk industrial cargo – Ruwais was developed as Abu Dhabi's major industrial port, also with substantial input by Gibb.

A breakwater to the east, and another to the south-west, were added in 1978 and 1979, using a sand core armoured with rock from Ras al-Khaimah, 150 km away. They are armoured with 3-m^3, 7.5-t concrete tetrapods.

The original lighterage wharf was built of 40 cylindrical cells of straight-web steel-sheet piling filled with sand. Subsequent deepwater berths have a precast concrete deck supported on cylindrical steel piles. The berths for harbour craft constructed in the second phase are of tied back sheet pile construction and the dhow harbour is constructed from precast concrete blocks.

There have been some problems of concrete deterioration. 'The codes and design methods of the 1970s were not tough enough to produce fully durable concrete,' says Peter Hunter. 'We've now installed cathodic protection to extend the life of some structures substantially.' Modern practice would use one or more techniques, such as denser concrete, a higher cement content, epoxy-coated reinforcement, cathodic protection, or the addition of slag, ash or micro-silica to the concrete to extend its life. Mass concrete is also used much more: 'The cost of producing high quality reinforced concrete increases the desirability of mass concrete structures. A lot more concrete is needed but the structure has a longer life,' says Hunter. 'Complex reinforced concrete structures don't have an indefinite life in the Gulf.'

Sewerage

In 1974 five British companies and one French firm were invited by the ruler to take over responsibility for the Abu Dhabi sewerage project. The existing consultant had fallen out of favour as a result of smell problems at the treatment works. John Taylor & Sons was the successful bidder.

Taylor produced an interim report on Abu Dhabi in early 1975 and a master plan later the same year. Interim works enlarged the system to cope with the expected increase in flows up to the point when the master plan works were commissioned.

Construction of the master plan works, worth US$1,100 million to date, began in 1977 and these were commissioned in stages, beginning five years later. Works included over 1,000 km of GRP and PVC sewers and GRP pumping mains, in diameters of up to 2.2 m and at depths of up to 10 m, plus 20 major pumping stations. A large treatment works included preliminary, primary, secondary and tertiary treatment. As in Dubai at about the same time, Abu Dhabi embarked on a major programme of effluent reuse for irrigation. In addition, sludge is digested, dried and bagged as a fertiliser.

A complete system, including pumping stations and mains, and storage reservoirs, has also been built to distribute the treated effluent throughout the conurbation. Tertiary treatment for the effluent includes rapid gravity sand filtration, and chlorination or ozonisation to disinfect it. The effluent is used both for agricultural and municipal irrigation, but the emphasis is on beautification. Ultimately 70 million m^3 of treated sewage will be used annually in the conurbation. The irrigation system is designed to prevent people coming into contact with the treated effluent. A mixture of direct underground drip-feed, sprinklers, trickle irrigation and hand-held hose application is used on a total area of 5,000 ha in the city of Abu Dhabi itself. This area comprises public gardens, highway roundabouts, verges and central reservations, and specially-planted 'amenity forests'.

Elsewhere in the emirate, D. Balfour & Sons (now part of the Maunsell Group) was responsible for design and supervision of sewerage for al-Ain, which was expected to grow to a population of 185,000 by 1995. Since 1981 when phase I began, US$75 million has been spent on laying 10 km of trunk sewers, construction of a sewage pumping station and oxygen injection plant, laying 15.5 km of ductile iron pumping mains and 15.5 km of GRP rising main. A sewage treatment works is capable of handling 27,000 m^3/d with provision for a doubling of capacity. Tertiary sand filters and chlorination produce effluent which is suitable for irrigating parks and roadside verges.

Buildings

Gibb had what former chairman, the late Geoffrey Coates, described as 'a very strong hold' on work in Abu Dhabi, similar to the privileged position enjoyed by Halcrow in Dubai. Both Gibb in the UK and Gibb Petermuller in Athens were involved in a wide range of schemes.

In 1976 the presidential court became concerned about a serious shortage of hotels in the emirate and decided to institute a programme of hotel building, calling for four hotels in six years. Gibb was appointed consultant on the project.

To gain time, the client decided that the hotels should be designed, built and equipped on a turnkey basis since this would allow finalisation of design, and preparation of interior design, to continue into the construction period. Gibb was to act as project manager and assess the designs.

The client brought in hotel operators at an early stage so that their requirements and standards could be incorporated into the design. This did not always work because the operators' requirements could not always be satisfied within the turnkey contractors' budget, which was already set within the fixed-price contract. However, three of the four were completed to budget and programme.

In total, 1,000 beds were provided by the four hotels, which were the Ramada, Meridien and Sheraton, all in Abu Dhabi, and the Ramada at Jebel Dhanna, 250 km west of Abu Dhabi. All were innovative in some way, at least in terms of construction practice in the Gulf.

The Ramada Abu Dhabi was the first to be built, being completed in 1978. The two-storey building has a steel frame clad with a dry wall system. The Meridien also has a structural steel frame with precast, pre-stressed concrete floor units manufactured at Jebel Ali. The West German contractors working on the Sheraton used fast-track methods to erect a reinforced concrete frame, part precast and part constructed in-situ by slipforming. The Ramada, Jebel Dhanna, was built on the beach and vibro-consolidation was used to improve the bearing capacity of the foundations. All these techniques were new to the area at the time.

Simultaneously the emirate wanted to start a programme of hospital building. As Geoffrey Coates recalled: 'Three or four panels of doctors had been set up to try to get consensus over what Abu Dhabi needed in the way of medical care, training facilities and so on.' When this failed to produce results the presidential court decided to go for package deals. Gibb was given three weeks to assess the somewhat vague proposals that came in. Since this essentially meant assessing the whole of medical services provision for the emirate it was an almost impossible task. 'But somehow we managed to get things on to more or less an even keel,' said Coates. 'All we could do was report on the proposals which had come in and point out what was missing.' The work was handled by Gibb Petermuller & Partners in Athens, and the consultant was retained to assess the turnkey contractors' designs, manage the projects and undertake site supervision.

The programme comprised the Zayed military hospital, al-Jazeerah general hospital, Tawam hospital in Al Ain, Mafraq, and the Corniche maternity hospital. They were all built between 1976 and 1984 at an overall cost of US$224 million and equipped to the highest standards.

Consulting engineer Jan Bobrowski & Partners has a reputation for working on concrete stadia and was responsible for several such structures in Abu Dhabi. The most prestigious was the grandstand for the Abu Dhabi National Day Parade completed in 1983. The 287-m long structure comprises two stands to the east and west, flanking a central VIP pavilion. The central pavilion is constructed in in-situ concrete, with its floors, terrace and beams post-tensioned with unbonded tendons. The east and west stands are identical, and are primarily of precast construction. Concrete strengths at 28 days of 50 N/mm^2 and 60 N/mm^2 were called for. No previous design in the United Arab Emirates had called for concrete as strong as 60 N/mm^2 and there was some doubt about whether it could be achieved with the locally available aggregates and workforce. However, according to Bobrowski partner B. K. Bardhan-Roy, the South Korean contractor Thaiwan had little difficulty achieving this. In fact, for the precast concrete, an even higher strength of 70 N/mm^2 was achieved.

Using the experience of the National Day Parade stand, 60 N/mm^2 concrete was again specified in the al-Wathba camel-racing stand, for the pretensioned roof beams which support the translucent polycarbonate sheeting of the structure's barrel vault shell. These beams cantilever 12 m over the seats at the front. The main structure consists of two rows of reinforced concrete arches 10 m apart running the length of the stand, spanning 18 m. The al-Wathba stand was completed in 1982, ahead of the National Day Parade stand where the fixing of lavish finishes added considerably to the contract period. Costain was the contractor.

The main stand at Al Ain football stadium, built a year later, is unusual in that its roof structure, supporting arches and VIP enclosure are separate from the seating and standing area below. Like al-Wathba, the main structure consists of two rows of in-situ arches, this time 6 m apart and spanning 24 m. At each end a 'half arch' cantilevers 12 m. The roof beams are made of precast concrete which was post-tensioned before erection. They are supported by the arches and cantilever 15 m over the seating. A thin slab follows the soffit of the arches and connects the two rows together.

The 1991 al-Maqam horse-racing grandstand at Al Ain is essentially a royal box for 300 people, with three entertaining rooms for VIPs and their guests, each with its own terrace. Built almost entirely of reinforced concrete, it again has an arcaded main structure. This time the arches span 18 m, supporting 25-m precast post-tensioned roof beams cantilevering 25 m.

Airports

Abu Dhabi has two international airports, one serving Abu Dhabi city itself, and one serving Al Ain. The original Abu Dhabi city airport was adjacent to the city on Abu Dhabi island, and approaching its design capacity of 1.5 million passengers a year by the early 1970s. In 1974 Aéroports de Paris, the French state organisation which built and operates the Paris airports, was appointed consultant to advise on an extension.

It concluded that there was limited scope for developing the existing airport for projected passenger flows to the turn of the century and so a study was started for a new airport.

The result, on a 12-sq km site 37 km from downtown Abu Dhabi, was built between 1977 and 1982 when it was opened. It was designed for 3 million passengers a year with scope for this number to double by the end of the 1980s.

When Aéroports de Paris was appointed Charles de Gaulle airport in Paris had recently been opened, and the new Abu Dhabi airport shared the same architect, Paul Andreau, as well as the concept of a 'satellite' terminal, a circular building which in Abu Dhabi's case has 11 telescopic gangways. The terminal building, its satellite, the control tower and the state reception room were built by Takenaka Komuten and Kumagai-Gumi of Japan. Other contracts covered the power house, runways, a freight terminal and other ancillary buildings. Innovative features of the £90 million project include fuel hydrants and electrical power supplies sunk into the apron, to cut down on the number of vehicles needing to use the apron.

At the same time, plans were also being developed for another international airport at the emirate's second city, Al Ain, on the mainland 150 km east. Al Ain is at an altitude of 250 m, and so has a more temperate climate. It was being developed as a centre for tourism and agriculture and is also the location of the United Arab Emirates University.

Scott Wilson Kirkpatrick was appointed consultant to the project in 1978. Its first task was to find a site for the airport because a location originally identified in the early 1970s had been encroached on by proposed development of the town. It was also considered too close to Omani airspace. The new airport required an area of 6 km x 2.5 km, with 25-km radius of airspace around it. The runway was to be 4 km long with 300-m overruns at each end, and a take-off slope of 1 per cent was to be provided. Early investigations concentrated on an area of desert 35 km away and near the Al Ain–Abu Dhabi road, until the client added an additional requirement, that the site must be within 15 km or 20 minutes' travel time of Al Ain. This eliminated desert floor sites and the search then concentrated on areas of sand dune ridges. A location was eventually found just within 15 km which satisfied all the critieria. However, 25 million m^3 of earthworks would be needed to level the site.

The master plan was developed for this site, with the runway orientated almost north/south, cutting across the dunes which ran north-east/south-west. The plan is designed to incorporate very long-term expansion, with generous development areas provided around buildings for later extensions. A new 9-km wide dual carriageway was built from Al Ain to provide access to the airport.

The earthworks required to level the dunes were the most challenging aspect of the project, representing one of the largest single earthmoving operations in the Middle East. The runway cut across two major lines of dunes rising to over 100 m. Elsewhere the dunes rose to 25 m above their troughs. The dune formation was ancient and therefore stable. All built-up areas were kept above the surrounding terrain to make it less likely that drifting sand would settle on the airport where it would cause operational and other problems. Further protection against this was achieved by laying a 100-mm blanket of protective gravel, quarried locally, over all exposed sand surfaces. Sand fences to prevent sand drifting on to the airport were considered but shown by trials to be unnecessary. In its natural state the sand was under-compacted: measures were taken to ensure that airport traffic did not cause it to compact. Surprisingly, the possibility of flooding and erosion under occasional but intense rainfall also had to be dealt with.[1]

The runways and aprons are designed to take the largest civil and military aircraft. There are three aprons, for passengers, for a future Royal Pavilion, and a freight terminal. Ultimately the passenger terminal is intended to have segregated arrival and departure lines and will be capable of taking 700 passengers hourly.

The work was carried out under separate contracts for the earthworks, which alone took three years; pavements and services; general buildings; furniture and fittings; power supplies and ground lighting; and air traffic control equipment. Inauguration of the airport took place on 31 March 1994.

Water supply

British engineer Peter Hipwell was seconded by John Taylor & Sons to Abu Dhabi as director of water distribution for Abu Dhabi between 1975 and 1979. Over that period a complete new supply system was installed. 'When I got there it wasn't very good,' says Hipwell. 'We rebuilt it totally.' The existing system dated from the 1950s, he estimates, and could not cope with demand. As a result the island of Abu Dhabi had been divided into zones, and each zone received water for two to three hours a day. An army of workers went round turning valves on and off.

The new system was designed by the Dutch consultant Tebodin, with modifications where necessary by Hipwell and his staff. It consisted of a 1,200-mm ring main running right round the island, pressurised by two pumps. All the water was desalinated, and was provided by two desalination plants working on the multi-stage flash evaporation system, each with an output of 23 million l/d. The principal contractor for the rebuilding programme was Harbert & Howard of the US. At the end of the project the consumption of the 250,000 people on the island was 140,000 m^3/d. Service reservoirs with a storage capacity of 360,000 m^3/d were also included in the scheme.

Peter Hipwell and his staff were also responsible for water-supply systems elsewhere in the emirate. On the island of Abu al-Abyad, west of Abu Dhabi, water was tapped in the desert and conveyed by a 900-mm underground pipe 95 km to the shore. But it had to go through an area of shifting dunes. 'The dunes kept blowing away and exposing the pipe so that it collapsed,' says Hipwell. A solution was eventually found, based on the reasoning that: 'The dunes only move because of the wind blowing. If you cover the pipe with material which the wind can't shift, the dunes move over the pipe, while the pipe stays where it is.'

In 1988 D. Balfour & Sons was awarded the design and supervision of a major project to carry 75 million gallons/day of desalinated water from Taweelah to Al Alin town, 150 km away. The project involved laying two large diameter pipelines, pumping stations, control systems and so on. Construction started in 1994 and will cost US$475 million.

D u b a i

Sewerage

In 1965, well before the discovery of oil, development was already gathering pace in Dubai. Import and export trade was thriving. The population was around 60,000, almost all concentrated around the mouth of the 12-km tidal creek where the twin trading towns of Dubai and Deira are sited opposite one another. The old towns covered an area of 3.8 sq km around the original port and were densely built up, with narrow streets. More modern development covered a further 12 sq km; this was more spacious and provided with wide roads.

The introduction of piped water had made existing sewage disposal facilities totally inadequate, and J. D. & D. M. Watson (later

Watson Hawksley and now Montgomery Watson) was called in at the direct request of the ruler to report on the problem. The firm has been working on Dubai's sewers ever since, enjoying for many years the same close relationship with the ruler on sewerage development as Halcrow had for ports and heavy civil engineeering work. 'Until five years ago the Watson office was like an extension of the municipality,' says Montgomery Watson engineer Alan Willson, who spent 15 years working in Dubai. 'If anyone wanted to know anything about sewers in Dubai, they came to the Watson office.' This lasted until the late 1980s when more consultants became accredited and competition was introduced.

Not only was Watson one of the first British firms to be established in Dubai but one of its engineers can also claim to have coined the name 'United Arab Emirates'. Watson engineers were in those days prominent as volunteers on the local radio station, Dubai Broadcasting, which broadcast throughout the Gulf in English. Watson Hawksley engineer Mike Thompson was scheduled to read the news bulletin one night a few months after the United Arab Emirates had been formed. 'The director of the station happened to come into the newsroom and said that there was a problem with the name. At that time they were using the name "Union of Arabian Emirates", but they didn't like it. I asked what the Arabic was it was "Al Emirat Arabia Intacta" which literally meant "United Arab Emirates", so I said, "why don't you call it that?" so we put it out on that night's news bulletin – it was agreed with Sheikh Rashid's office – and the name stuck.'

In 1965, the consultant had studied Dubai's sewage disposal arrangements and concluded: 'Provision of a sewerage system to convey waste products safely away from the more densely populated areas of the town is now a matter of urgency.'[2]

The story is familiar from elsewhere in the Gulf. Drainage from older parts of the town was to soakage pits, while newer properties had cesspools. The soakage pits were leaking into the ground and polluting the creek; the cesspools were not emptied regularly and also overflowed. To make things worse, there was a high groundwater table, within 0.6 m of the surface in places. In other words, the town was a public health time bomb.

One of the consultant's headaches was deciding what population size to plan for. At that time oil had not been discovered and Watson decided that the population might ultimately reach 100,000. It proposed in the first instance to provide main sewerage for 45,000, and to construct a treatment works to serve 25,000. Phase I was constructed between 1968 and 1971, giving Dubai the first fully operational modern sewerage and sewage treatment facility in the Gulf.

Phase II received the go-ahead in 1972, by which time it was clear that the population could significantly exceed 100,000. A major review was carried out in 1976 leading to the phase III report. This made proposals for dealing with a population of more than 100,000 and also considered options for effluent reuse. Beautification using treated effluent to irrigate extensive planting has become one of the characteristic features of Dubai.

Because the topography is flat, considerable pumping of sewage is needed. In the first two phases, 100 local pumping stations were built. In addition there are two main pumping stations on the Deira side and three on the Dubai side. Conditions in the sewers are highly septic. The first sewers were constructed using epoxy-lined asbestos cement pipes but these were superseded by corrosion resistant unplasticised polyvinyl chloride (uPVC) and GRP, the GRP pipes being made locally.

All the work in phases I and II was carried out by the Orient Trading & Contracting Company (OTAC), which laid 400 km of sewer from 100-mm house connections to 1,000-mm mains in the first two phases. Construction was particularly difficult in the narrow streets around the souks of the old town, where the sewers had to be constructed through hundreds of cesspools and soakage pits, which had to be emptied, disinfected and demolished as pipelaying progressed. Wellpoint dewatering was used to allow excavation, given the high water-table.

The major feat of the first two phases, however, was the construction of a triple pipe crossing the creek. Two 750-mm diameter pipes and one of 375-mm were laid to carry sewage to the treatment works on the Dubai side, and to allow for the possibility of pumping treated effluent back. The crossing was constructed by closing off half the creek at a time and excavating through sand bunds.

The original sewage works was a conventional biological treatment plant, with pre-aeration to reduce septicity. It was extended in 1976–9 to cope with a population of 100,000. From 1978, tertiary treatment, consisting of rapid gravity sand filtration and chlorination, allowed 5,700 m^3/d of effluent to be reused, some of which went to the Dubai cement works. More importantly, however, the availability of treated effluent for irrigation allowed the ruler to achieve the aim of creating a park in the city. Following the successful establishment of

Saffa park, irrigation was extended from 1982 to roadside verges and central reservations, transforming the appearance of the city and belying its desert location. Creation of golf-courses and racing tracks for both horses and camels was also made possible.

A completely separate distribution mains network was provided to convey the treated effluent around the city. One of the most striking features of this non-potable water system is the Jumeirah water-storage tower, 36 m high with a capacity of 1,000 m^3. Dubai municipality wanted to produce a prestigious structure, and the tower is decorated with a pattern reflecting traditional Islamic designs, cast into the concrete and painted in contrasting brown and cream.

The city continued to expand until by the early 1980s it had reached the site of the treatment works which was originally 6 km into the desert. This limited scope for expansion. A new site was identified 14 km from the centre and a new treatment works there, built to a German design, came on stream in 1993, superseding the original one.

Other consultants and contractors became involved in later stages of the scheme. Acer Middle East, for example, was commissioned to upgrade and rehabilitate 50 km of sewers in Deira in 1990 to ensure both the structural integrity of the sewers and their capacity. Closed-circuit TV inspection was used to check the condition of the system and computer modelling to identify sewers which were either surcharged or which had substandard velocities.

Ports

Dubai's development in the 1970s was epitomised by a series of increasingly spectacular and ambitious schemes by which the ruler, Sheikh Rashid, sought to use his newly acquired oil wealth to ensure the emirate's long-term economic prosperity. Thus a new deepwater harbour, Port Rashid, was followed by the massive dry dock project, which itself was eclipsed by the huge industrial port of Mina Jebel Ali, excavated out of dry land on the shore.

Dubai during this period was refreshingly free of the bureaucracy which afflicted many other Gulf states. Instead, the ruler placed strong reliance on personal relationships as the basis for awarding commissions and took a close personal interest in their progress. Sir William Halcrow & Partners had become established in Dubai at an early stage. The close and trusting relationship which developed between Sheikh Rashid, and Neville Allen, the Halcrow resident representative from

1955 to 1977, gave the firm almost complete dominance of civil engineering consultancy in the emirate for many years. As Halcrow's 125th anniversary brochure puts it: 'The Ruler entrusted to Halcrow most of the Emirate's civil engineering projects and paperwork was kept to a minimum. All approval was verbal, a far cry from the cut-throat post-war arena of Europe.'[3]

The ruler's style is effectively illustrated by the story of the inception of Mina Jebel Ali, quoted by the same source. A Halcrow representative, Bill Briggs, was summoned to an isolated desert spot to meet the ruler. Surrounded by mile upon mile of nothing but sand, the engineer wondered what the reason for the summons could be. Soon it became clear. 'The ruler pointed to a vast area and described his concept of what would be one of the largest harbour developments in the world. There was one small hurdle. As the ruler anticipated a stormy meeting with some influential but restive merchants that coming Tuesday, work would have to start on site the preceding Monday morning. . . .'

Work did indeed begin on the appointed day on this project, which, at a cost of £835 million, was at the time one of the largest construction projects ever undertaken.

Halcrow's involvement had begun much more modestly, however. In the 1950s, Dubai creek was inhabited by a lively population of traders, pearl-fishers and gold smugglers. It was the centre of the dhow trade with Bombay and Basra. The ruler became concerned when the creek began silting up resulting in damage to trade. Sheikh Rashid borrowed funds to improve the creek and, because of the strong links between the emirates and Britain through the maritime truces, turned to the British government to suggest a consultant which could help. Halcrow's was the name put forward, because the firm had an office nearby in Kuwait. The consultant proposed a scheme which kept the creek open, and this formed the basis of the close relationship which developed.

In 1965 Halcrow was commissioned to report on the feasibility of the first of the ruler's huge maritime schemes, the £160 million Port Rashid. The consultant then designed the project and supervised its construction. Trade had grown to the point where the shallow creek could no longer cope, and a deepwater harbour was called for. Phase I of the construction programme began in 1969. It included 15 deepwater berths, an oil tanker berth, a transshipment basin, 9 ha of storage, plus roads and other associated infrastructure. When inaugurated in 1972, Port Rashid was the largest deepwater harbour in the Middle East.

It has been said that the number of berths was increased to fifteen, from the four which the consultant originally considered justified, at Sheikh Rashid's insistence. If so, the ruler was proved right because the port was soon working at full capacity. Only four years after its inauguration, work began on phase II, to provide a further 22 berths. This required the lengthening of one of the original Stabit-armoured rubble-mound breakwaters and the construction of another to increase the sheltered area of water from 150 ha to 450 ha. Breakwater construction used 3.5 million m^3 of rock and 25,000 precast Stabits, weighing between 7 t and 15 t, as armour.

Most of the phase I berths and all those of phase II were 180 m long and dredged to a depth of 11.5 m. The original berths had been designed for general cargo; five of the phase II berths were designed for container freight, and two of the original berths were also strengthened and converted for containers. A further two berths handled ro-ro traffic. The original port continued to operate at full capacity while the extension was built. It was opened in 1980.

Construction of Dubai's dry dock was in many ways the logical next step after Port Rashid. The original idea for a dry dock had come from OAPEC which proposed the scheme to enable the Gulf to benefit from shipping as well as oil. The Gulf is ideally located for carrying out repairs to tankers. Before any work can be carried out, their tanks have to be cleaned and filled with inert gas. This is normally done on the voyage back to the loading terminal. Dry docks in Europe are too close to the point to which the oil is delivered for the tanks to be cleaned before arrival. A dry dock within a day's sailing time of the loading terminal, however, is ideal because it allows tank cleaning to be completed *en route*; then, after repairs have been done, proving trials can be carried out between the repair yard and the terminal.

Sheikh Rashid commissioned Halcrow to carry out a feasibility study for a dry dock capable of holding a 250,000-dwt tanker, and the consultant produced a favourable report in 1971. OAPEC, meanwhile, decided to site its proposed dry dock – the Arab Ship Repair Yard – in Bahrain. However, the US promoter and consultant John J. McMullen suggested to the ruler that a bigger dry dock would be a viable proposition in any case. The UK construction companies Taylor Woodrow and Costain were brought into the discussions.

The ruler saw that this would fit into his strategy of seeking security for Dubai's economy in the long term. McMullen, Taylor Woodrow and Costain worked together on further feasibility studies. The ruler decided to go ahead and set up the Dubai Dry Dock Company, in which McMullen was a director. Halcrow was brought in once again as consulting engineer.

The plan was to build not one, but three, dry docks, of which the biggest would be capable of holding 1 million-dwt tankers – which were at that time on the drawing board, but in fact have never been built. The dock would be built out from the shore adjacent to Port Rashid, minimising disruption outside the site. The dock was to be capable of carrying out a wide range of general repairs, and, since not all repairs need the ship to be out of the water, there were also to be a number of ordinary berths. Even by the standards of the Gulf, this was an enormous undertaking.

Four years were initially allowed for construction and a fast-track approach was adopted, with design and construction taking place simultaneously. An unusual contractual arrangement, a hybrid of design-and-construct with the conventional system of design and supervision by the consultant, was used. Although the consultant retained ultimate responsibility for the design of the works, the contractor was responsible for producing detailed designs, which were agreed step by step with the consultant.

This provided flexibility for work to begin on site as soon as the contract was signed, allowing mobilisation and a start to be made on dredging, breakwater construction, and reclamation. Everything else was initially priced as a provisional sum in the contract, with a requirement for a definitive priced design to be agreed within six months of the contract start.

An ICE Proceedings paper on the dry dock in 1979 said that this procedure speeded up decisions on detail design criteria. Other benefits were it said that: 'Difficult construction details were minimised and both the designers and construction management firmly believed that the agreed concepts could be put into practice.' This system also provided flexibility for making substantial variations to the design as the contract progressed.[4]

Taylor Woodrow's head of special projects, Chris Irwin-Childs, was involved in the dry dock from start to finish. He says: 'We wouldn't have been comfortable dealing with a scheme we did not understand. All sorts of problems had to be overcome which under a conventional form of contract would have been insoluble.'

The three docks measure respectively 370 m x 66 m for vessels of up to 350,000 dwt, 415 m x 80 m for vessels of up to 500,000 dwt, and 525 m x 100 m for vessels of up to 1 million dwt. They can be dewatered within four hours, and the gates can be raised or lowered within 20 minutes. They are grouped with the largest dock in the centre, with pump houses and heavy-lift cranes serving two adjacent docks.

It was decided to build the dock walls of reinforced concrete caissons cast on shore, launched by shiplift and floated into position. They are self-buoyant, extend the full width of the dock piers and are filled with sand. A total of 162 caissons was produced in the two years from May 1975, generally measuring 31 m across, 17 m in length and 17 m in height and weighing 3,500 t. Slipforming was used to cast them and the two gantries in the production yard produced an average of 1.7 caissons a week. It took 20 days to make and cure a single caisson.

Aggregate for reinforced concrete was brought from the village of Haha, 110 km inland. An 80-km access road was built which at the ruler's request was upgraded and designed to full highway standards to form part of the Dubai–Oman highway.

For floating into position each caisson was ballasted with an initial 1,000 t of sand. Up to 8 m was dredged from the sea bed to give a firm rock foundation at a depth of 14.2 m. A trench either side was accurately drilled to a depth of 1 m, and 1-m downstands on each caisson fitted into these trenches. Once in position, the void beneath the caisson was grouted. This arrangement was designed to minimise diver work.

The caissons were then filled with sand. This raised the question of what pressure to design the caisson walls to resist, since it was not clear how the sand fill would behave within the caisson cells. Initially, a conservative approach based on Rankine theory was adopted. However, tests on site showed that the fill, well-graded granular material tipped by lorry into the cells, was not compacting. The consultancy firm Ove Arup, which had been employed by the contractor to undertake the geotechnical aspects of the dock design, warned that there was a danger of excessive settlement being caused by vibration or shock loading. It was decided to vibrate the fill to cause it to settle, but this could have potentially caused high pressures and overstressed the caisson walls.

An elaborate test programme was adopted to devise the best method of compaction. Settlements of up to 900 mm were induced using a 12-m steel tube poker. Monitoring confirmed that pressures on the walls were reasonably accurately predicted by Janssen's method of design for silos.

The dock floors are formed of 1.5-m thick reinforced concrete slabs, with a gravel drainage blanket beneath leading to open-jointed drains leading to pumps, to prevent build-up of pressure under the floor. The slab is designed to allow a ship to be placed anywhere within the dock.

Ships enter the harbour, which is protected by breakwaters totalling 4.1 km in length, under tug control. They are pulled into and out of the dry docks by a specially designed system of four winches, two at the head of the dock and two at the entrance.

Dock 2's enormous width of 100 m presented serious problems for the design of a gate, if this was to be designed to act as a horizontal girder across the dock. Instead, the contractor proposed a vertical propped cantilever supported by the sill and by a hinged prop part way up the gate, a design which allowed the weight of steel to be kept to a minimum. The gates are hinged at the bottom outside the dock and, when open, both the gates and props lie in recesses below sill level. The gates are formed of 4.1-m wide buoyancy modules alternating with stiffened plates and were floated as 37 modules from the fabricators Redpath Dorman Long/Clarke Chapman in the UK. The design, known as the Promod system and developed from Taylor Woodrow's concept by the consultant T. F. Burns & Partners, was patented.

The gates are raised and lowered by winches which have to lift a load of only 8 t. The design is such that, when closed, all hydrostatic forces are taken by thrust pads, separate from the hinges which therefore have to carry only the self-weight of the gate and props. A synthetic rubber seal keeps the gates watertight, and a 10-m deep cut-off wall below the sill prevents seepage underneath.

The dry dock is comprehensively equipped to allow almost any repair to be carried out: a decision to extend the onshore repair facilities led to an extension of the contract beyond the original four years. There are 70 onshore structures, including administration blocks, a steel fabrication shop, a machine shop, maintenance yard, pipework and galvanising shop and so on. Equipment in the workshops includes lathes of up to 23.5-m bed length to undertake turbine repairs. A workforce of over 3,000 is employed.

The dry docks themselves have cranes of a maximum capacity of 120 t at 65.5-m radius, which with their jibs raised are taller than St Paul's Cathedral in London. One of the crucial tasks for a ship repair yard is to repaint ships as rapidly as possible, and at Dubai an automated, hydraulically operated system ensures that this can be done in four days for the largest ships.

The dry dock was officially opened by Sheikh Rashid and Queen Elizabeth II in February 1979. Although it has been a commercial success, the reinforced concrete structure has suffered considerable deterioration in the adverse conditions of the Gulf, and discussions are still going on to determine how to deal with this.

Even by the standard which Port Rashid and the dry dock had set, Sheikh Rashid's next major project, the industrial port of Mina Jebel Ali, was hugely ambitious. Situated south of the creek near the border with Abu Dhabi, the 66-berth port to accommodate 40,000-dwt ships was conceived as the heart of a new industrial area whose population was expected to grow to 300,000 by the end of the century. It required 3.3 km of breakwater, and dredging and excavation of 110 million t of rock, including the provision of a 16-m deep approach channel 17 km out to sea.

One of the most remarkable features of the development – said to have been suggested by the ruler himself – was that the 470-ha basin was carved largely out of the desert. This was not the most economic engineering solution, but provides better access to the berths than the more usual construction method in which deepwater berths are served by a causeway. Sheikh Rashid believed that the inland construction would be a major asset in encouraging industry to locate there.

Industry set up there from the outset included an aluminium smelter built by British Smelter Constructions, of which Wimpey was part; a steel fabrication works built by Trafalgar House and Cleveland Bridge; a cable factory for Dubai Cables (a joint venture between BICC and the ruler); a power station, a liquefied petroleum gas plant and an hotel.

The requirement for a quick start to be made and for a berth dredged to 8 m to be operational within 18 months, had the contractor clamouring for drawings before a site investigation had been completed. To get the work under way quickly, the first quay was constructed using 1-m diameter steel pipe piles driven to a depth of 18 m in pre-bored holes in the rock, anchored by tension rods to a tieback wall near the surface. This method was adopted because pipe piles were readily available and the rock was suitable for auger boring. It allowed construction of the first quay to begin before the quay wall design had been finalised.

The design which was eventually adopted for the quay wall employed huge unreinforced concrete blocks laid on top of each other, without mortar – essentially the same method as that by which the pyramids were constructed 3,000 years ago. Each block measured 6 m x 2 m x 2–2.2 m deep and weighed 42–45 t. The cost of this method of construction was similar to that of other methods, but the blocks could be laid equally easily on dry land or under water.

Tight dimensional tolerances were called for, with each pair of faces required to be parallel to within 3 mm. A £2 million concreting and block-making plant made the blocks at a rate of 100 a day in order to keep up with the rate of excavation. The production line and storage area was 2 km long.

The blocks contained no reinforcement, not even any lifting eyes. This meant that sea water could be used in the concrete mix. Halcrow stipulated maximum chloride and sulphate contents of 2.5 per cent and 4 per cent respectively. Salt water was also used in the 31,000 Stabits. Fresh water had to be used, of course, in the precast capping beams which are reinforced.

Because the blocks had no lifting hooks, they were handled in the block-making yard by a grab or clamp which simply gripped the block on either side. For final positioning, a special clamp which engaged in a groove along the ends of the blocks was used. A total of 15 km of quay wall was constructed using 65,000 blocks.

The port has two basins, the outer partly dredged and partly excavated to a depth of 14 m and the inner excavated to 11.5 m. Wellpoint dewatering was successfully used to keep the excavations dry despite a groundwater-table only 1 m deep. In total 110 million m^3 of sand, sandstone and breccia was dredged or excavated. The main contractor for the dredging was Gulf-Cobla, a joint venture between the Dubai Transport Company and Costain-Blankevoort. Some of the approach channel was sublet to the Great Lakes Dredge & Dock Company of Chicago.

The dry excavation in the inner basin was subcontracted to a joint venture, Dutco-Pauling, which mobilised one of the biggest fleets of plant ever seen, to shift the 31 million m^3 of earth comprised in its contract. The fleet included twenty Caterpillar D8s, nine Cat D9s, nineteen Terex TS24s, seven O&K backhoes and sixty Euclid R50 dump trucks.

The main civil engineering contractor was Mina Jebel Ali Construction, a joint venture grouping Dutco, Balfour Beatty, and Stevin Middle East (the latter being a subsidiary of the Dutch Stevin group). Al-Futtaim Wimpey supplied rock for the breakwaters and built a necessary diversion of the Abu Dhabi–Dubai main road.

The 235-m wide approach channel was initially intended to be only 8 m deep but part way through the contract the required depth was doubled, so that it had to be taken 17 km out to sea. Gulf-Cobla used four cutter-suction dredgers built for the project for all except the hardest areas of rock, which had to be blasted. Material was pumped ashore through floating pipelines for reclamation. To get the channel down to the increased depth an additional heavy-duty cutter-suction machine had to be specially ordered from Mitsubishi of Japan at a cost of £10 million. A total labour force of 4,500 was employed on site, all from overseas.

Conceived at the same time as Mina Jebel Ali and built between 1977 and 1981, Dubai's aluminium smelter, costing US$1,300 million was one of the largest industrial projects ever undertaken. The original concept was developed by Sheikh Rashid and Paul Brauner, then chairman of British Smelter Constructions which was at that time 50 per cent owned by Wimpey, the latter having been responsible for the Bahrain smelter in 1968.[5]

Several circumstances combined to make it the right project for Dubai at the time. First, the ruler controlled the source and price of gas which would provide power for the smelter and which normally accounts for 30 per cent of an aluminium smelter's operating costs. Second, it was clear that the development of Dubai, and the creation of the industrial zone at Jebel Ali, would leave Dubai with a serious water shortage which could not be met from natural sources. If a desalination plant were built at the same time as the smelter, waste heat from the smelter's power plant could be used to produce steam for the desalination plant at minimal cost. A third reason which made the smelter attractive for Dubai, with its small population and need to rely largely on immigrant labour, was that, once operational, the smelter would need only a small workforce. So the decision was taken to build a smelter to produce 135,000 t/y of aluminium (compared with 120,000 t/y at the Bahrain smelter), with a 515-MW power plant and a 25 million g/d desalination plant using the flash evaporation process.

British Smelter Constructions was managing contractor for the 52-month turnkey contract, with Wimpey doing most of the civil engineering work. Funds were provided by the UK's Export Credits Guarantee Department and the German Export Credit agency. The overall engineer was P. U. Fisher supported by consultant Kennedy & Donkin.

The contract was split into eight sections: site preparations and administrative buildings; raw materials handling; power; the reduction plant; the anode plant; the cast house; workshops and repair shops; and the desalination plant. Site preparation and administrative buildings included accommodation for the workforce which peaked at 4,000 expatriate workers. For raw materials handling a dedicated berth was provided at Jebel Ali, 6 km away, with mechanised unloading equipment and storage silos capable of handling 5,000 t of coke and 30,000 t of bauxite.

Power for the plant is provided by a 515-MW gas turbine power station which burns gas from an offshore field. Conversion of bauxite to aluminium takes place in the reduction plant which houses 360 cells, or pots, in three lines. The anode plant produces the anodes, made from blended petroleum coke fractions and baked in kilns, which are consumed at the rate of nearly 100,000 t/y. Molten aluminium is made into ingots or billets in the cast house. Substantial workshops and repair shops are provided because there were none available locally.

The eighth part of the scheme is the desalination plant, in which sea water is heated by steam produced from waste heat from the power station, and passed through a series of chambers each at a lower pressure and temperature than the last, in which vapour condenses as fresh water. Of the 25 million g/d produced, the smelter needs 500,000 g/d.

All three pot lines were producing aluminium three months ahead of schedule in 1981. Eight years later Wimpey won another, two-year, turnkey contract to add a fourth pot line, increasing production capacity to 240,000 t/y.

A more recent port development in Dubai has concerned the creek, where dhows still tie up. Congestion was becoming a problem there as up to 700 vessels jostled for space. Maunsell & Partners won a design competition to provide more dhow wharfage, relieving congestion and encouraging further development of trading by dhow. The scheme provided four basins separated by finger piers, with moorings for up to 250 dhows and other vessels. The quay walls, of which there are 3,250 m, are built from precast mass concrete blocks each weighing 30–50 t. Administration buildings, service roads, car parking and cargo storage are also provided, and because of the basin's proximity to the city centre, considerable attention was paid to landscaping.

The scheme was opened in May 1993 by the ruler. During the construction of this project, Maunsell was asked to design the nearby Bani Yas road extension, a dual three-lane road with an underpass and a tunnel. Maunsell supervised the construction of this project which was opened in 1994.

In 1993 Maunsell was commissioned to carry out feasiblity studies for the development of the Al Mamzar area around the lagoon and the Port Saeed area adjacent to the Dubai creek. Currently the consultant is designing the infrastructure for the above areas in addition to the creek sea walls.

The 39 storey Dubai International Trade and Exhibition Centre was the tallest building in the Gulf when it was built in the late 1970s. Five Gulf states can be seen from the 37th level viewing gallery; the top two floors contain TV broadcast facilities. Floor 33 houses the restaurant, and the lower floors contain offices and a 170 seat lecture theatre.

The building has a reinforced concrete central core for lifts, stairs, service ducts, and toilet and kitchen facilities. Reinforced concrete plate floors span between the core and the perimeter. Deep column sections provide sunshading, which, together with the orientation of the building, help to minimise the air-conditioning load. The building is founded on a 2-m thick pile cap supported by 776 piles founded on well compacted sandstone.

The complex also comprises a four storey exhibition hall whose main hall provides 2,500 sq m of space, the six storey Dubai Hilton hotel, and three 14–15 storey serviced apartment blocks.

Architect for the project was the John R. Harris & Partners, consulting engineer was R. J. Crocker & Partners and the contractor was Bernard Sunley & Sons.

Water desalination in the Gulf

Demand for water continued to increase in the United Arab Emirates as development progressed. In the late 1980s Sir M. MacDonald & Partners (later Mott MacDonald) was commissioned to oversee a programme of installation of eight desalination plants at different Gulf locations, using the relatively new technique of reverse osmosis.

Distillation as a technique for obtaining fresh water from a saline source had been known for centuries, but it was not used on a large scale until the 1950s with the growth of oil developments in the Middle East and the Caribbean. At this time the development of multi-stage flash distillation revolutionised the process and produced an economic and relatively trouble free method of supplying drinking-water. It remains the most economic method for large-scale use.

Reverse osmosis was introduced in the 1960s and developed rapidly. Originally it was only suitable for brackish waters (that is water with a salt content less than that of the sea) but has since been developed to be suitable for sea water also. 'Reverse osmosis is much cheaper than distillation in small plants,' says Mott MacDonald director Mike Burley. The maximum practical capacity of an individual unit is around 45 million l/d.

By the early 1970s two of the largest reverse osmosis plants had been built in Saudi Arabia: a 46 million l/d plant at Salbukh treating brackish water, and a 12.5 million l/d sea water plant at Jeddah.

Osmosis is the phenomenon by which, when two solutions of different concentrations are separated by a semi-permeable membrane, solvent will flow from the dilute solution to the more concentrated one. A pressure known as the osmotic pressure is associated with this.

In reverse osmosis a pressure greater than the osmotic pressure is applied to cause water to flow from the more concentrated to the more dilute solution. Some 95–98 per cent of the salt can be removed in a single pass. In fact rates of rejection higher than 95 per cent may not be needed since a low level of residual dissolved solids is desirable in drinking-water. A disadvantage of distillation is that the water it produces is too pure so has to be blended with brackish water for human conusmption.

Development of reverse osmosis partly depended on the production of better membranes. Cellulose acetate was the original membrane material used in the 1960s. In 1970 Du Pont introduced polyamide fibre membranes, whose hollow fibres are no thicker than a human hair. In the late 1970s composite materials began to be introduced, in which a thin film of polyamide or other substance forming the membrane itself is supported on a porous substrate of a different polymer which offers support.

Polyamide hollow fibres are self-supporting but the other types need a suitable structure to allow them to withstand the high pressures at which reverse osmosis plants operate. A plant treating brackish water would typically operate at, 1,400–4,200 kN/m^2, while for sea water pressures would range from 5,600 kN/m^2 to 7,000 kN/m^2.

There are limitations of temperature and pH ranges in which membranes can operate. Because water in the Middle East can come out of the ground at temperatures of up to 55°C, it may have to be cooled. Membranes are also prone to clogging and fouling if the plant is not operated carefully. Pretreatment of the water, such as softening or acid dosing, may be needed to prevent calcium carbonate or sulphate scale precipitating on the membrane. Cellulose acetate membranes can be affected by biological fouling which can be prevented by chlorination.[6]

Sir M. MacDonald's commission in the late 1980s for eight sea water desalination plants around the Gulf was one of the biggest single programmes for installation of reverse osmosis plants. They are located in Jebel Dhanna, an island off Abu Dhabi; Ajman; Fujairah; Sila; Mirfa, Abu Dhabi; Dhabaia; Rafeek; and Sur, in Oman. These were all remote places or islands where no alternative supply existed. The consultant prepared tender documents, evaluated the turnkey bids and approved the contractor's detailed designs and supervised construction. The plants produce 1,100–9,000 m^3/d of permeated water from raw sea water which has a salinity of 40,000–48,000 ppm and temperatures in the range 20–35°C. The overall cost of the eight plants was £27.5 million at 1987 prices.

Roads in Abu Dhabi and Dubai

Both Dubai and Abu Dhabi had extensive road-building programmes during the 1970s and 1980s. Freeman Fox & Partners worked in both emirates. In particular, in Dubai it became in effect a branch of the municipality, in a position analagous to that of Watson Hawksley for sewerage, says former Freeman Fox & Partners senior partner Derek Wolstenholme.

The biggest project was the 115-km Dubai to Al Ain road. This was designed and built as a dual two-lane carriageway with grade-separated interchanges. It passed through an area of shifting dunes so those near the road were sprayed with bituminous material to inhibit them from moving.

Later, in 1978, Freeman Fox & Partners upgraded 23 km of the road between Abu Dhabi and Al Ain to dual three lane expressway standard. The road carried heavy traffic because it provided a strategic link to Oman and oil and gas fields in the Gulf. Before overlaying and reconstructing the road, Freeman Fox & Partners carried out extensive investigations to examine areas where the construction had broken down. Where extensive reconstruction was needed, the removed material was incorporated into the sub-base of the new carriageway.

Freeman Fox & Partners was also extensively involved in construction of smaller town roads in Dubai, as well as in a one-off project for 50 km of local roads in Al Ain. Its work in Dubai continues to the present day.

As part of a proposal in the early 1970s for a coastal highway, a tunnel across the mouth of Dubai creek was needed. The Deira–Shindagah tunnel, as it is known, was designed by Halcrow which also provided construction supervision. The 300-m cut-and-cover tunnel cost £9 million. It consists of a reinforced concrete box divided into two 9.2 m compartments each containing a two-lane carriageway. A third compartment on one side provides a pedestrian subway with a services compartment below.

It was constructed within sheet-steel pile cofferdams, in two stages to allow the creek to remain open. Axial flow fans provide longitudinal ventilation. An advanced level of services was provided, with automated control of ventilation, lighting, and traffic controls.

Sharjah

Sharjah's oil resources are relatively modest compared with those of Abu Dhabi and Dubai and development there has been on a smaller scale. Nevertheless, like its neighbours, the emirate built its own deep-water port and international airport.

A close relationship, similar to that between Halcrow representatives and the Ruler of Dubai, also existed between the Ruler of Sharjah and Halcrow representatives John Greenhalgh (between 1968–1977) and Mel Stewart (1977 to date) and helped the company win work in Sharjah throughout this period.

Ports
Between 1972 and 1978 Halcrow designed and supervised the construction of Sharjah's Port Khalid, a £60 million deepwater port with 15 berths protected by 5 km of breakwater. Later, in 1976, Khor Fakkan container port, on the Indian Ocean coast, followed. This was sited in a narrow bay with natural deep water, minimising the amount of dredging needed. The port has a 430-m quay with water dredged to 12.2 m; the two container berths are constructed of blockwork (confirming the trend away from reinforced concrete now that its vulnerability in the aggressive conditions of the Gulf is recognised), and are protected by a 565-m stabit armoured breakwater. There is an 8-ha back-up yard with space for stacking 3,000 containers.

Airports
A £50 million international airport for Sharjah was commissioned by the ruler in 1974. Facilities provided include a passenger terminal, air traffic control building, VIP lounge, a cargo building and the control tower. The runway is 3,760 m long to cater for Boeing 747s.

Siting of the airport, 15 km from Sharjah town, took into account the emirate's boundaries and the fact that Dubai's international airport is only 20 km away. The scale of the terminal's passenger throughput, a relatively modest 400–500 passengers an hour, underlines the fact that the emirates were sometimes driven more by the desire not to be out-done by their neighbours rather than by awareness of what they strictly needed.

The ruler insisted that the terminal should reflect Islamic art, and the result is a striking complex of three domed buildings. The airport opened in 1979. Since then the consultant, Halcrow, has worked on extensions.

Sewerage
Provision of main drainage in Sharjah followed the pattern of that in the other emirates. Since 1974, capital works worth US$350 million have been built, providing foul sewerage, sewage treatment works and surface water drainage system. There is a distribution network for treated effluent which is sand filtered, chlorinated, and used for irrigation. As in Dubai, the flat topography requires a large number of pumping stations. Oxygen injection is used in the long pumping mains to reduce septicity. Phase I was commissioned in 1977; phase II, providing capability to deal with a population of 75,000, was commissioned in 1983. The con-sultants were Halcrow with D. Balfour & Sons (now part of the Mannsell Group).

The UK construction company Tarmac was responsible for laying 36 km of pipes for phase IA and IB, for constructing two pumping stations and the sewage treatment works in a three-year, US$37 million con-tract. With a workforce of 650–700 at peak, Tarmac laid up to 1 km of pipe a week. After the problem of the rapid deterioration of concrete pipes became apparent, pipes made of asbestos cement lined with coal tar epoxy, concrete with plastic liners, and, in phase IB, vitrified clay, were adopted. Pipes already laid were lined with GRP.

The contractor also built 12 km of 600-mm diameter fresh water pumping mains from oases in the desert to the city of Sharjah in 1978–9. It carried out £3.5 million of asphalting at Sharjah airport, and built the 46-km Sharjah–Dhaid dual carriageway costing US$27 mil-lion, plus a US$6 million interchange for access to the airport. It also won the civil engineering work for Sharjah's US$24 million power sta-tion, built between 1978 and 1982.

Buildings

When the ruler commissioned a new souk (market), which was built in 1982–7, he insisted that the design should incorporate features drawn from both Islamic art and classical eastern culture. He specified that it was to be built in stone which he said should resemble that of the Natural History Museum in London. The result, designed by the Halcrow Group's architectural practice, is a structure 200 m long. Two 90-m long vaulted shopping arcades two storeys high meet under a central tower, where refreshment areas are housed under the 14.4-m diameter central golden dome.

The architecture explores the theme of light which in Islamic art is seen as the perfect symbol of divine unity. The dome's interior depicts the stars in Sharjah's sky as they would have appeared on the night the ruler was born, with a pole star centre-piece 3 m across made up of several thousand separate pieces of glass. The cost of the souk was £8 million.

The Northern Emirates

Having little or no oil resources, the other emirates have had to rely on receiving a share of oil income from their neighbours to fund development. However, Ras al-Khaimah and Fujairah turned out to have a different type of natural asset: they are among the few places in the Gulf where good quality aggregate can be found. This was of course in huge demand when construction was at its peak.

Among quarrying developments there was the Fujarac aggregate project, for the Fujarac Rock & Aggregate Company, for which Ove Arup & Partners was project manager and consulting engineer. This provided a fully automated aggregate exporting facility with an annual production capacity of 3 million t of crushed rock. The contract included construction of the crushing facility itself, where substantial foundations were needed for the heavy machinery. Crushed rock is transported 3 km to a new berth at the port of Fujairah, for seagoing barges of up to 17,000 dwt. Construction of the berth, including dredging 450,000 m^3 of material, was achieved in six months.

The Emirates today

Consultants, particularly those engaged on water and sewerage schemes, have found it easier to maintain a presence in the United Arab Emirates than in many other parts of the Middle East, despite the 1980s' downturn in workloads (although some, like Halcrow, whose Gulf workload represented 95 per cent of annual revenue at its peak in 1979, were nonetheless severely affected when the downturn came). The same was not true of contractors, however, only a handful of whom have managed to maintain a presence.

One of the exceptions was Laing Emirates. Laing had been late in entering the United Arab Emirates market: its first job was construction of the 22-storey commercial–residential Deira tower in 1977. This led to a number of contracts for the royal family, including the camel-racing and horse-racing tracks. There were also two spectacular golf-courses, the Emirates course and the Creek course, which required much excavation and feature artificial fresh-water lakes, and are built to full international championship standards.

But Laing remained active during the lean years of the 1980s partly by dint of opening a block-making factory, but more importantly by starting a joinery business, Laing Emirates Joinery, in 1984. 'We perceived a need for high quality interiors in a market like Dubai,' says Laing's Dubai director Bruce Foot. The firm set up a factory under its control, able to achieve standards 'equivalent to anything in the UK'. This has paid off handsomely: the subsidiary now employs 150 workers, and recently won an order worth £1 million from the Dubai Meridien Hotel.

The emirates are seen as continuing to offer reasonably good prospects. Jim Robinson, Acer Consultants' Middle East managing director, said in mid-1993: 'All sorts of construction projects are in the pipeline.' These are mainly in Abu Dhabi, Dubai and Al Ain. They are, however, relatively small, though Acer has enough work to keep a staff of 200 in the Gulf. 'We are ticking over,' Robinson said, with competition 'very fierce' there. 'People are flocking back because of the world recession.' But, he added, 'there are few mega projects except in the petrochemical industry' where basic infrastructure is beginning to need renewing.

c h a p t e r   n i n e

OMAN

In the last two decades, Oman has undergone a remarkably rapid transformation even by the Gulf standards. When it began to produce oil in 1967 the country was impoverished and almost totally undeveloped. What had been a successful maritime trading state until a century before had been eclipsed in the age of the steamship, and had regressed to conditions which some historians have described as 'medieval'.

Sultan Said, who ruled the country at the time that oil production began, appeared unable to visualise, or to accept, the modernisation of his country which the people were demanding and oil development had made possible. When he was deposed in 1970, Oman had only 10 km of tarmac roads and three primary schools – yet over the next decade and a half, standards of living were transformed as an oil-based economy developed. Moreover, these changes were brought about with a remarkable absence of social tension.

Oman, the third largest country on the Arabian peninsula, stands at the gateway to the Gulf and overlooks the Strait of Hormuz through which most of the oil exported from the region travels. There are two main areas of population: the 'capital area' along the narrow coastal strip in the north around Muscat, backed by the Hajar mountains; and the Salalah plain in the south. Musandam, a rocky peninsula on the tip of the Strait of Hormuz, is separated from the rest of the country by a 70-km strip which is part of the United Arab Emirates. Historically, Oman derived its importance from its position as a stopping-off point on the merchant shipping route to Africa and India, and its inhabitants were traditionally seafarers.

Archaeological evidence of settlements dating from prehistoric times has been found. According to legend, Arabs first arrived around the sixth century AD, fleeing from Yemen after the collapse of the Marib dam there. Oman's history is characterised by conflict between the coastal population and that of the interior, and between tribes. The Arab population was converted to Islam in the mid seventh century AD. About half the population are Ibadi Muslims, a sect which believes in elected leaders, and is receptive to technological change. Women are encouraged to play a full part in the economic life of the country.

Oman was at various times in its history dominated by Persia, Portugal and Turkey. The present ruling dynasty dates from 1749 when it was founded by Ahmad ibn Said after he had driven out the Persians. Oman signed a treaty with Britain in 1839, allowing for a British representative to be stationed in Muscat. For a century from the mid-1850s Oman lapsed into relative obscurity as its maritime trade declined. In 1879 the basis of modern Oman was created when the southern province of Dhofar was merged with the rest of the country.

The first oil concessions were granted in the 1920s, but little progress was made because of the hostility of the population and the absence of roads. The country at this point was extremely poor. Sporadic fighting continued, for example in the 1950s, between the people of the interior and the coast. A long-running border dispute with Saudi Arabia over Buraimi broke out again at the same time. In 1965 there was an uprising in Dhofar in which the population protested about the country's lack of development. The uprising was aided and abetted by the Marxist regime in neighbouring South Yemen.

Oil was first discovered in 1950 but it was a heavy crude which was uneconomic to produce. In 1967 an economically viable oilfield was discovered. However, the ageing Sultan Said failed to capitalise on it and to embark on the modernisation programme which his people wanted. In 1970 he was overthrown by his son, Qaboos bin Said, who had been under house arrest for six years. Qaboos had been educated at the Royal Military Academy at Sandhurst in Britain.

Qaboos ended the civil war in the south partly with the help of the British army but also through a 'hearts and minds' campaign to show that he was committed to raising living standards. This removed the rationale for the uprising.

Through a series of five-year plans, beginning with a programme of schools- and hospital-building, the country was modernised. In the mid-1970s it appeared that the country's oil resources would last for only a few years. But by 1980 the price had risen enough to make production of the heavy oil from the south viable, and, in the breathing-space that this provided, substantial new discoveries made the outlook more promising.

Oil remains the basis of Oman's economy despite attempts to diversify into other industries. The country's first oil refinery was opened in 1984 at Mina al-Fahal. In the same year copper mining, which has a history in Oman dating back 5,000 years, restarted at Sohar.

From a construction point of view, the main problem of working in Oman was inaccessibility, at least before the road system was built. As elsewhere in the Gulf, high temperatures in Oman could cause difficulties in casting and curing concrete. In general, however, the environment is much less aggressive there than in many other Gulf countries. There are no problems with saline groundwater attacking concrete, for example. High-quality aggregates are available locally.

Buildings

When Sultan Qaboos came to power he took control of a country with only three primary schools and just 900 children in primary education. One of the new sultan's first acts was to instigate a programme for building six schools and hospitals. These were constructed by British contractor Costain, in joint venture with the Swedish company Skanska.

Improved education for the Omani people remained a policy commitment, culminating in the sultan's promise in his 1980 National Day speech to build a university by 1986. The university's aim was to give students an education which would prepare them for the modern world while maintaining Islamic and national traditions and identity.

Creation of a university in such a short time was a formidable challenge. A first step was to create a foundation committee under the chairmanship of the Minister of Education & Youth Affairs, which included a team of academics from Oman, the UK, the US, and Egypt. The committee decided on the academic form of the university; there were to be five colleges for medicine, engineering, science, agriculture, and education. A faculty of literature and the arts was to follow.

The committee also decided that the colleges should share facilities such as lecture and seminar rooms, the library, and a conference hall and cultural centre. Male and female students were to be kept separate, resulting in a two-tier system of walkways to allow women to circulate round the campus without being seen by men. It was estimated that two-thirds of students would need to live on site, and 13 residential blocks, each with 150 study bedrooms, were provided. The male and female residential blocks both had their own sports facilities. Residential accommodation for teachers was also provided. The university was planned to have an eventual complement of 10,000 "students.

The British architectural firm YRM was appointed by the client to develop the academic brief. January 1982 had already arrived by the time the architect was asked to prepare a master plan, but within three months a plan had been produced and approved by the sultan. YRM's partner in charge, Fred Woodhead, said later that the design evolved naturally, almost creating itself from the foundation committee's requirements.

Special facilities for individual colleges were to include a model farm for the agriculture college. The medical department was to have its own library and a teaching-hospital was planned to be ready by the time the first intake of students began clinical studies.

The UK construction group Cementation International was awarded a turnkey contract for the university project in March 1982, giving it just two months to mobilise and 48 months for detailed design and construction of the £225 million project. A turnkey contract was considered most appropriate, to allow planning and design to continue into the construction period. Partly for this reason, and partly to provide accommodation for the workforce, work began on the residential blocks for students and staff, the simplest buildings on the project.

Cementation also appointed YRM as its architect under the contract, and a system was set up under which problems were all referred upward to a high-level committee of three with authority to make rapid decisions. This consisted of Mike Slater, director in charge for Cementation, Fred Woodhead of YRM, and the university vice-chancellor Sheikh Amor bin Ameir. This arrangement was an essential factor in meeting the deadlines, Mike Slater told the UK journal *Construction News* in 1987: 'When anyone had difficulties at any level, instead of the problem stalling at that level and precious time being wasted, it was speedily raised to the level where it could be addressed and solved.'[1]

The tight timetable focused everyone's minds; from the client's side there was complete commitment to make the scheme work. Woodhead paid tribute to Sheikh Amor, for holding together the project and ensuring that there were no trivial changes of mind.

Design and construction, comprising 250,000 m^2 of buildings in an area of 3 sq km, were divided into seven packages: university staff housing; student residences; colleges; support facilities; infrastructure and landscaping; and furniture, fittings and equipment. Trafalgar House Engineering Services was the civil and structural subconsultant to YRM.

To speed construction repetitive elements were used wherever possible. However, there was insufficient time to mobilise for extensive precasting of major elements. In-situ reinforced concrete was the chosen material for the university buildings. Limited use of secondary finishes also helped to cut down construction time. One of Fred Woodhead's great pleasures in the job was, he said 'the immediacy . . . drawings would leave the office on a Friday night and we would be seeing the construction arising from those drawings before the end of the following week.'

YRM selected the site, 40 km from Muscat and 8 km from Seeb international airport, from three options. The chosen site has a central valley area which was used for the main teaching buildings and the most striking building on the campus, the mosque. The valley lies east–west, oriented towards Mecca, and the mosque is placed at the end of the main east–west axis through the complex.

However, the university site lies within the bed of a wadi and was therefore vulnerable to flooding in the rare but torrential rain that falls in this part of Oman. For flood protection the central buildings were founded on a platform of fill. A mound in the valley had to be removed first, requiring the blasting away of 150,000 t of rock; to create the platform, 1 million t of earth had to be moved. In addition, there is an extensive stormwater drainage system, and a feature wadi which runs through the site.

YRM's buildings are designed simply and incorporate typical features of Omani architecture. A recurring motif is the use of flat arches with cruciform columns. The common facilities buildings are arranged along a formal avenue running from the main gates to the mosque and are generally more lavishly finished than the college buildings which surround them. There are four 150-seat lecture theatres and another, of 450 seats, housed in an octagonal building. The library is faced in red sandstone with large stained-glass windows on two facades. The mosque dominates the site, however. It has a gold-coloured dome and twin minarets, and is clad in pale pink sandstone with decoration in blue, turquoise and gold leaf tiles.

Significant infrastructure work was associated with the university. A total of 8 km of roads was built, including a 4-km orbital route. There is a sewage treatment plant capable of coping with a population of 10,000. Mains for water, and natural gas – of which the university was the first domestic user – were also required. The workforce, mainly Indian and Thai, reached over 4,000.

The first students, 320 men and 200 women, were able to take up their places in September 1986 as planned – a tribute to the considerable efforts of the entire project team.

An ambitious public health policy was also part of Sultan Qaboos' philosophy and the royal hospital serving the capital area, built at around the same time as Sultan Qaboos University, was central to this policy.

The hospital was built also as a turnkey project by Wimpey International, which won the contract in July 1983 and assembled a financial package for it. The hospital has a gross floor area of 64,000 m^2 and is built on a 25-ha hilltop site. All branches of medicine are covered; there are 629 beds as well as outpatient and emergency facilities. It has over 2,000 personnel, and accommodation is provided on site for 700 medical and support staff. Because of the Omani tradition for patients to be accompanied by their families, there is also extensive visitors' accommodation.

Percy Thomas Partnership was appointed architect under the contract and Donald Smith, Seymour & Rooley was building services consultant. Wimpey drew on its subsidiaries for structural engineering, and supply of mechanical and electrical services. Wimpey Alawi, its Omani subsidiary, began work on site in December 1983.

Blasting and excavation of 40,000 m^3 of material was needed to prepare the site. The hospital buildings consist of a reinforced concrete frame on a 7.2-m grid with a storey height of 4.5-m. This includes a 1.2-m deep service zone within the false ceiling. Concrete, of which 42,000 m^3 was used, was produced mainly on site at a peak rate of 200 m^3/d. Visitors are provided with separate perimeter corridors to the wards, to minimise disruption to the hospital's work.

Oman's Ministry of Health specified the most advanced medical technology to equip the hospital. For example, X-rays and other test results can be transmitted instantly around the hospital by computer. There are six operating theatres including one for open-heart surgery and one for orthopaedic surgery.

As with Sultan Qaboos University, there is considerable infrastructure associated with the hospital. There are separate roads for ambulances, service supplies and visitors, and the hospital has its own sewage treatment plant. Recycled effluent is used to irrigate the landscaped gardens.

In the days of the former sultan, there were few opportunities for construction firms from abroad. One company which did work in the sultanate before the *coup d'état*, however, was the contractor Taylor Woodrow. The company went into Oman in 1969 and has worked extensively there ever since in association with the local firm Towell. Its first two projects were a small hospital – the first in the country – and houses for Petroleum Development Oman. One of its most prestigious later jobs was the Sultan Qaboos Sports Complex, built to celebrate the 15th anniversary of the sultan's accession to power. The complex includes a stadium, built to Olympic standard and accommodating 30,000 spectators, together with a sports centre, football and hockey pitches, athletics fields, and an outdoor Olympic swimming pool. It was designed by architect Roy Lancaster Associates and engineer Coode International.

Taylor Woodrow-Towell was also part of a consortium which developed the new residential area of Medinat Qaboos. This provided accommodation for 5,000 people in the Qurm development in a project which extended over 15 years.

Ports

As oil revenue began to flow in the late 1960s, Sultan Said started to consider the country's economic development, albeit on a limited scale. As in other Gulf countries, it was soon obvious that port facilities were inadequate, and Sir William Halcrow & Partners was appointed to advise on the development of the harbour at Mutrah. The consultant was asked to prepare first a master plan with recommendations for phased implementation of the project. On approval of the master plan, contract documents were to be prepared at once to allow construction of the first phase to go ahead as soon as possible.

This comprised two lighter berths capable of handling 150,000 t of cargo annually without congestion, with transit sheds, a port administration building and passenger facilities able to deal with 100 passengers an hour.

In March 1970 the first-stage contract was awarded to the Belgian, Dutch, French and Spanish consortium, Six Construct. The two lighter berths totalled 300 m in length, with a minimum water depth of 4.3 m. A rock-armoured breakwater 460 m long was to protect the berths. Provision was made for the mooring of small tankers delivering gas oil to a nearby oil company depot. The £2.1 million contract was completed by 1972.

Halcrow's Viv Hoad, who went to work on the port a few months before the coup in 1970, found things 'hot and rugged' and restrictive. Sunglasses and shorts were banned and the sultan's permission was needed to use a vehicle.

Originally the master plan had envisaged a slow development of the port, to be carried out in four stages, to a capacity of 1 million t of cargo by 1980. Later extensions would allow the capacity to be increased to 1.5 million t. However, when Sultan Qaboos came to power, he gave instructions for the full development to go ahead at once. The harbour was renamed 'Port Qaboos'. Halcrow was instructed to negotiate with Hochtief, and this West German contractor was brought in as a joint venture partner with Six Construct on the expanded project.

This contract included a 700-m rubble-mound breakwater armoured with 15-t stabits; and eight berths totalling 1,524 m in length with water depth varying from 7.3 m to 14.6 m. One of the berths, 230 m long, could be adapted to take container traffic. A 17-ha storage area was created by land reclamation, with six transit sheds, workshops, and so on.

Plans were expanded further in 1973, when it was decided to add another deepwater berth 180 m long, together with a 76-m small craft berth and a 150-m dhow berth. An area was also to be provided for fishing boats. It was also decided to dredge the harbour basin to a minimum depth of 9.75 m, which required the removal of 460,000 m³ of material. The total value of the project to build the whole port was £21.6 million.

The quay walls are built of precast concrete blocks founded on a bed of rockfill in a trench dredged in the sand, or in one case on in-situ concrete in a trench in the rock. The blocks are a maximum of 1.7 m x 1.5 m x 5 m and were lifted by a special T-headed clamp which engaged in slots cast in the blocks. The quays are capped with an in-situ concrete slab which includes a services gallery and provision for future crane rail tracks. Eight of the ten berths have a transit shed placed centrally. The main breakwater has a heart of rubble protected by a layer of 1–6 t rockfill, obtained from the adjacent headland.

Roads

'After the old Sultan had been deposed, one of the first things they wanted was to build roads,' says Brian Brent, a director of Sir Alexander Gibb & Partners. As already mentioned, the Sultanate had only 10 km of metalled roads in 1970. Economic development of large areas of the country would obviously have been impossible without road access.

'In the first instance they arranged design and construct contracts with contractors,' Brent says. 'After one major contract had been let, the German contractor suggested that an engineer was needed to look after

above
Al-Fatah University, Tripoli,
is designed to accommodate 10,000 students on a 246 ha site.
© Ove Arup & Partners

left and below
In 1970, Oman had only 10 km of metalled road and less than 100 vehicles. The Muscat-Mutrah new road forms part of the thousands of kilometres of highway which have been built, opening up the country for economic development.
© Ozzie Newcombe

above
Roseires dam, on the Blue Nile, dramatically increased
Sudan's potential for growing crops through irrigation.
© Sir Alexander Gibb & Partners

top and above
The port of Ras Lanuf in Libya serves an industrial complex including
an oil refinery and petrochemicals plant.
© Rendel Palmer & Tritton

above
Libya's Great Man-Made River is claimed to be the world's biggest
civil engineering project and carries desert groundwater hundreds of kilometres
to centres of population on the Mediterranean coast.
© Sir Alexander Gibb & Partners

above
Prestressed concrete cylinder pipes of up to 4 m diameter carry the water.
© Sir Alexander Gibb & Partners

left
The Greater Cairo Wastewater Project: sinking the 32-m deep, 45-m diameter concrete caisson for the main pumping station at Ameria.
© Taylor Binnie & Partners

below
Tunnellers could erect a complete tunnel ring, consisting of seven concrete segments, in 40 minutes. A secondary brick lining was added later to protect against corrosion.
© Taylor Binnie & Partners

right
Extracting potash for fertiliser from the Dead Sea by evaporation is a major economic activity in Jordan.
© Wimpey Group Services Ltd

below
Evaporation pans can be clearly seen at the bottom of this satellite photo.
© Sir Alexander Gibb & Partners

above
The second Bosporus bridge, built in only two and a half years,
links Turkey to the trans-European motorway network.
© NKK Corporation Japan

the client's interests.' A shortlist was drawn up from which Gibb was chosen to review the design and supervise construction of 230 km of road along the north coast from Mutrah to Sohar. This was subsequently extended by another 100 km.

As a result of its work on this project, Gibb went on to take responsibility for design or supervision of 2,000 km of road, making up the bulk of the country's highway network.

A 140-km link from Seeb to Nizwa, built between 1973 and 1975, was the first road into the interior of the country, and began the process of opening it up for development. Seeb, on the coast west of the capital, was later the site of the Sultanate's international airport. Nizwa had previously been accessible only by four-wheel drive vehicles.

From Nizwa the network was extended 132 km north-west to Ibri in 1977–1980. In 1978 work began on building a 780-m link running the length of the country, to Thumrait in the south. Gibb was responsible for detailed design of the former, and for reviewing and substantially revising an existing design for the latter, as well as for supervising construction of both. The two roads were single carriageway – no more was justified by potential traffic growth at the time – with a design speed of 100 km/h.

'The terrain through which these roads were constructed was generally rolling desert plains,' says Brian Brent. 'But wadis were a major obstacle.' Rainfall is rare in the area but when it occurs it falls in substantial quantities, filling up the wadis or dry river beds. 'Very substantial flash flooding can happen at no notice. Providing a cost-effective solution to drainage was one of the major difficulties of the highways in the area.'

In general the solution was to use 'Irish crossings' (sometimes called 'Irish bridges'). This is a strengthened section of carriageway in the wadi bed, protected from the full force of the flood flow but designed to be submerged without being washed away. It is carefully designed to be safe to cross even in a flood, although not necessarily at peak flows, when it might be submerged beneath a metre of water. But its geometry should be such that the road clears quickly afterwards, becoming passable soon after the peak.

Design of an Irish crossing also aims to make it easy to maintain, with only a small amount of silt remaining behind and needing to be cleared after a flood. Concrete cut-off walls upstream and downstream are provided to protect the bed from erosion. 'A difficulty we faced was that there was very little historical data on flood flows, or even on rainfall – just a few years in Nizwa,' says Brian Brent.

Gibb was also involved in roads in the cities of Muscat and Mutrah, as well as the highway linking the two. Though they are only a few kilometres apart they are separated by a high mountain. Initially the existing mountain road was upgraded, but within a few years traffic growth rendered this inadequate. 'When we first went into Oman there were less than 100 vehicles,' Brent says. 'We were not at all surprised when the road needed dualling.'

Several feasible mountain routes were identified. 'We also looked at the coast, however, and decided that a road on reclaimed land around the edge of the sea was the most appropriate solution. All the reasonably level land in the area was already occupied. In constructing a corniche road we could arrange to reclaim a significant piece of land for subsequent redevelopment.'

The 3-km long dual carriageway, built between 1975 and 1979, cuts across two bays and provided 11 ha of reclaimed land. Some of this was on the seaward side of the road and is mainly used for parking and recreational use, but most was on the inland side. One major reclaimed area was a small bay which the road crossed; elsewhere, borrow areas created by blasting the mountainside to gain reclamation material were left terraced for subsequent development. Blasting the volcanic rock produced a durable, easily handled, well-graded material but not in sizes big enough for armouring sea defence works. Instead concrete armour units were used. Landscaping and planting to a high standard, plus an irrigation system, were included in the design.

The Seeb–Nizwa road was also affected by significant traffic growth, partly because of developments inland which the road itself had made possible, and in 1982 Gibb was commissioned to update it by adding a second carriageway. While keeping as much as possible of the original road, the new design upgraded the highway to a limited access road with major intersections grade-separated, adding service roads for local traffic.

In the Qurm development area west of Mutrah, Gibb, working as Gibb Petermuller & Partners, had a role which extended to provision of general infrastructure and planning. Qurm is a 3,200-ha site allowing for expansion of the capital area and is intended ultimately to have a population of 100,000; this represented a doubling of the population of the capital area and an increase in area by a factor of five. Gibb Petermuller was commissioned in 1974 to plan, design and supervise construction of the infrastructure, including roads and supply of other services. The contract also included the outline plans for two townships.

Provision was made for housing, light industry, and recreation, health and community facilities. There is a new diplomatic city for 45 embassies, with several government and ministerial buildings nearby.

In the south of the Sultanate, particularly along the south coast, road building presented a different challenge. There, mountainous and rugged terrain, coupled with a complete lack of existing access to the more remote areas, posed considerable problems.

The consultant Maunsell & Partners won the commission in 1981 to design 120 km of road between Mughsayl and Dhalqut, forming the final section of a link between the main southern town of Salalah and coastal villages at Rakhyut and Dhalqut. A joint venture between Balfour Beatty Construction and Oman Building and Contracting Company built the 78 km long first section of the road, from Mughsayl to the army outpost of Furious, between 1985 and 1990. The remainder has yet to be completed.

The mountain range in this region rises to 1,000 m above sea level, and the area is heavily incised by wadis with steep-sided valleys, often several hundred metres deep. The two-lane road has a 7-m wide carriageway with gradients limited to 10 per cent in general but with an occasional absolute maximum slope of 12.5 per cent. On the steep gradients the hard shoulder becomes a crawler lane. Hairpin bends are used where necessary on steep sidelong ascents.

The worst problems for the joint venture on its £51 million contract were posed by the 500-m ascent from the Wadi Afal valley. Added to the inaccessibility and the difficulty of getting any plant or labour to where it was needed was the weather, which included extreme heat and dense mists which could remain for months.

The project had three sections. Westwards from Mughsayl these were: a 12-km length through rock cuttings; a 5-km hairpin ascent from the wadi at Afal; and 61-km along the top of the escarpment to Furious. But to travel between the two base camps, one at Mughsayl and the other in the west at Arfit, necessitated a 320-km trek along desert tracks.

A total of almost 7 million m^3 of rock had to be removed by blasting, and the fissured limestone above the road had to be stabilised by 13,000 rock anchors, bolts and dowels. Elsewhere rock-faces were supported by shotcrete and overhangs supported by mass concrete buttresses. Fissures were sealed by grouting to protect anchors and bolts from corrosion. The joint venture was responsible for proposing stabilisation measures from a range of techniques, in the light of conditions on site, for the consultant's approval.

Most of the materials needed – aggregate for the concrete and asphalt, and excavated rock for fill – were available locally, but there was a severe shortage of water. When a 910-m deep well drilled on site failed to provide enough, Maunsell agreed to the use of sea water for compaction of embankments. This was delivered from the coast into a storage reservoir on site via a 13 km pipeline, and pumped up in stages from sea level to the 1,000 m high ridge.

The Afal ascent was always expected to be the most difficult aspect of the job and so it proved. It took the joint venture 18 months to establish a temporary route up the escarpment. At one time rock stabilisation put work six months behind schedule.

Stabilisation work continued at Afal for a year after the road as a whole was awarded its certificate of practical completion, until Maunsell was satisfied that the rock-face had been sufficiently restrained.

Work on the other sections went more smoothly. Though substantial rock cuttings had to be formed in the easternmost section, the work posed no unusual problems. The two largest wadis are traversed by Irish crossings rather than viaducts, the largest of these being 800 m long.

Expected traffic flow on the remote road was expected to be no more than 500 vehicles a day, though a high proportion would be heavy goods vehicles.

Maunsell was also responsible for the design of interior roads to several important agricultural villages in the Hafer mountains, 150 km from Muscat. During the early 1980s a 7.5-m wide single carriageway road over 190 km long was built to link Izki, Khadra Bin Daffa, Sanaw, Mudabyi, Samad and al-Haim to Oman's trunk road network. Link roads 7 m wide and built to a lower geometric standard served Mahleya, Seema and al-Hama.

The new road, costing US$160 million, generally followed existing tracks and wadi beds, across terrain consisting of gravel plains and rolling hills. Erosion protection was given considerable attention, with culverts and wadi crossings forming a major part of the works. It also crossed a number of ancient *falaj*, or irrigation tunnels. These were reinforced or reconstructed under the line of the road to ensure the continuation of the water supply they provided.

In 1990 Gibb was commissioned to carry out a national road transport development plan for the Sultanate from a strategic point of view. This involved an engineering and economic appraisal of 38 new roads and road improvement schemes and the ranking of them in order of priority in a programme to be implemented between 1991 and 1995. Traffic forecasts took into account the potential for traffic generation; social and economic development effects were also assessed.

Water supply and sewerage

Water supply played a key part in Sultan Qaboos's 'hearts and minds' campaign to end the civil war in south Oman in the years immediately after he came to power. Contractors, assisted by British army personnel seconded from the Royal Engineers, visited villages in the south carrying the government flag and sinking wells. This provided a reliable water supply for the first time in many villages where water had hitherto been distributed by carriage for long distances in donkey panniers.

In 1977 the World Bank, which had arranged a study of power and urban water supply, appointed Sir M. MacDonald & Partners to advise. The consultant carried out a comprehensive review of groundwater for 80 catchments, and of existing and projected future use. It considered the needs of the capital area and 30 other towns and villages.

In the capital area a water-supply and distribution system was needed to meet demand from 300,000 people by 1995. Sites for wellfields were identified to exploit groundwater resources, which were to be used in conjunction with an existing sea water desalination plant, which was also to be expanded.

Construction work has included 300 km of transmission and distribution pipelines, 128,000 m^3 of storage in eight storage reservoirs, water-towers, three major pumping stations, and a blending plant to mix desalinated water with groundwater. The distribution system covers 500 sq km. Ductile iron pipes are used for the transmission mains, and locally made asbestos cement pipes for the distribution system.

Sir M. MacDonald was also responsible for water-supply projects in four other major towns: Sur, Sohar, Nizwa and Buraimi. Since 1990, Watson Hawksley has worked on water and wastewater master plans for Nizwa, Sur, Buraimi and six other towns in northern Oman in association with MacDonald, as part of the latest five-year plan for developing the country's infrastructure outside the capital area. Populations now range from 7,800 in the smallest town, Saham, to 34,800 in Sur. These are set to double by the year 2010. At present less than 20 per cent of their combined populations is connected to a piped sewerage system.

During the 1980s water shortages and groundwater salinity became cause for concern in Oman. In 1988 the Ministries of Environment & Water Resources and of Agriculture & Fisheries were given powers to control groundwater resources countrywide. Sir M. MacDonald, assisted by Watson Hawksley, prepared a water resources master plan for the Sultanate to assist in future management, conservation and development of resources.

The firm also participated in a number of unusual schemes to recharge aquifers by diverting flood water. The first of these was at Khasab, situated at the confluence of three wadis on the Musandam peninsula. Three dams were to be built to protect the town from flooding and the idea of using the impounded water to recharge the aquifer arose in the course of design. The scheme has been successfully put into operation.

In the south of the Sultanate, John Taylor & Sons was responsible for developing groundwater resources for Salalah and Raysut. An ultimate population of 100,000 was designed for. A new wellfield of 10 production boreholes was developed, from which water could be pumped to meet an average daily demand of 32,000 m^3, feeding the two towns via storage reservoirs with capacities of between 3,000 m^3 and 27,000 m^3. Over 50 km of pipelines has been laid. The system is controlled by radio telemetry, and operation of the wellfield is regulated by microprocessor control.

Defence

Despite Sultan Qaboos's policy of combating the civil war and guerilla incursions from South Yemen by developing the country rather than primarily by force, a high level of defence spending was considered necessary in the early years of his rule. Partly because of the Sultanate's strategic position overlooking the Strait of Hormuz, such spending has remained high, accounting in the late 1980s for up to 40 per cent of government outlay.

One of the most spectacular military developments was the Thumrait airbase in the south, which boasts a 4-km runway, one of the longest in the world. Thumrait is situated at the top of the escarpment north of Salalah. The monsoon season brings low cloud to the south of the country for long periods and an airbase outside the cloud area was needed. The long runway is needed to allow fully laden aircraft to take off in daytime temperatures of 45–50°C: because the air density is low, aircraft have to attain a higher take-off speed to become airborne.

The base was a triumph of fast-track construction, with the airstrip becoming operational in December 1974, only eight months after the project was awarded, despite its being in a war zone. Sir William Halcrow & Partners was the consultant and Paulings the civil contractor. The airstrip was made available to the Allies during the Gulf War. A workforce of 5,000 was assembled from India, Sudan and the UK. Cement had to be flown in from Abu Dhabi, and 1,000 km of track maintained for convoys. Halcrow director Mel Stewart says that the work in this part of the world could be particularly hazardous because South Yemeni guerillas were able to take advantage of the monsoon mist to lay mines. He recalls that, while he was working on roads near Salalah:

> The enemy would come out of the hills at night and lay their traps. Every morning we'd look for footprints to make sure no mines were buried under the area where we were working. I had a specially adapted Land-Rover fitted with seatbelts, and the floor and underseats packed with sandbags, so if you ran over a mine, it blew the wheels off but didn't blow you up or out of the vehicle.'[2]

Oman today

Since 1970 Sultan Qaboos has succeeded in transforming his country and in bringing about substantial improvements in living standards – changes which have come about with surprisingly few signs of social stress. Higher standards of living and the commitment to education have won the allegiance of the people of Oman and succeeded in defusing unrest in the south of the country.

Oman's oil reserves are relatively modest, however, and in common with most of the other Gulf states its long-term prosperity will depend on diversification of the Sultanate's economy. So far, efforts to encourage this have met with limited success.

YEMEN

For over 3,000 years frankincense and myrrh have played a major role in religious ceremonies. South-west Arabia, being one of the few places where the trees from which they are extracted grow, was thus important for many centuries as a supplier of these sought-after commodities. Sea and overland camel trade routes radiated from there to the contemporary centres of civilisation. The region is also a traditional source of Mocha coffee, named after the port from which it was exported.

This is Yemen, where in the 1960s the two Yemeni republics, which merged in 1990, were established. North Yemen, the more fortunate in terms of climate and mineral resources, had emerged from nominal Ottoman control after World War I to become an Imamate, while South Yemen is the former British Colony and Protectorate of Aden. The history of the modern republics has been one of internal and external conflict, with the brief unification which took effect in 1990 threatened by civil war in 1994.

With its high mountains and considerable rainfall, Yemen differs from the rest of Arabia, and for centuries was an important agricultural centre. Some 2,500 years ago the Sabaean kings built substantial buildings in dressed stone as well as extensive irrigation systems which incorporated canals, drainage ditches, tanks, sluices and tunnels, and the great dam at Marib which was held together by lead ties and copper nails. In use for about a thousand years, the dam was destroyed by a flood between AD 542 and AD 570, an event which is believed to have brought about the downfall of Yemeni civilisation at that time.

Unique in the world are the mud skyscrapers of Shiban in the Wadi Hadhramaut in south-east Yemen, which was a long-established trading centre and which for centuries had close ties with Muslim communities in Indonesia and other South-East Asian countries. Rising up 10 storeys and built entirely from mud and straw, they are an imposing sight and are reputed to be over 500 years old.

Tradition also has it that the castle of Ghumdan was erected by an Arab king in the first century AD to protect his townspeople against nomads, and that it had 20 storeys, each 10 cubits high, of granite, porphyry and marble.

North Yemen was conquered by Islamic armies in the seventh century AD. Later, in 1517, it was conquered by the Ottomans; they left in 1636, returning two centuries later. However, during the second period of Ottoman rule, residual authority in outlying areas of the country remained with Yemeni imams, who took over power after World War I. North Yemen was thus one of the first states in the area, together with Saudi Arabia, to achieve independence.

Imam Yahya, who ruled from 1904 to 1948, followed a policy of isolating the country from outside pressures and influence. The economy stagnated. In 1948 he was assassinated by the Free Yemenis, members of a movement which grew out of the opposition of merchants and intellectuals, but the uprising was defeated by the imam's son Ahmad. Another uprising was put down in 1955 but dissatisfaction continued to grow among Yemenis as they compared the poor economic conditions in Yemen with developments taking place elsewhere. On Ahmad's death in 1962 a group of army officers, influenced by the nationalist movement in Egypt, seized power and proclaimed the Yemen Arab Republic (YAR). Most Yemeni heads of state have been army officers since then.

The immediate result of the proclamation was to plunge the country into civil war. Ahmad's son Badr escaped to rally royalist opposition. Initially both sides received outside help: Egypt sent forces to support the republic, while Saudi Arabia supported the royalist faction. The war continued for eight years with neither side gaining the upper hand. It was eventually ended in 1970 with the establishment of a coalition which included many figures from the royalist side, though the former ruling family of Hamid al-Din remained exiled from the country.

This did not bring peace to the republic, however. Internal conflict continued. Between 1974 and 1978, three sucessive heads of state were deposed or assassinated. Meanwhile radical nationalists, who opposed the setting up of the coalition government, began a guerrilla war which escalated in 1976 with the formation of the National Democratic Front (NDF). However the regime of President Abdullah Saleh, which began in 1978, proved more stable. In 1982 a cease-fire with the NDF was arranged, bringing peace to the republic for the first time.

Since the cease-fire the YAR has had good relations with Saudi Arabia, which, having initially seen it as a threat, went on to provide aid. The republic has pursued a non-aligned policy in its relations with the East and West but relations with the West have been good, and economic aid has been provided from that source too.

The YAR's economy remained primarily agricultural until the discovery of oil in 1984 in the Marib region, near the border with Saudi Arabia and South Yemen. At that time agriculture accounted for 80 per cent of GDP, with coffee the chief cash export. Many Yemenis went to work in the oil-producing states, especially Saudi Arabia, and remittances provided a significant source of foreign exchange. Oil production is now about 200,000 b/d.

In the south the Sultan of Lahej broke away from the imam in the north in 1730, but it was only during the nineteenth century that South Yemen really emerged as a separate entity. Until then Aden, which had been a flourishing port since mediaeval times, had been integrated with the North, as had much of the interior. Only the Hadhramaut area, based on the fertile valley of the Wadi Hadhramaut in the east, had been autonomous, building up trade and political links with India, Singapore and Indonesia.

Aden was seized by Britain in 1839, on the pretext that the Sultan of Lahej had allowed piracy against British ships. Britain was alarmed that the ruler of Egypt, Muhammad Ali, had gained too much power among the states in Arabia. It saw this as a threat to the trade route to India, and occupied Aden to provide a secure coaling station for its steamships.

Aden developed as a British naval base, and though Britain had little interest in the interior, its colonial administration gradually expanded during the next century and culminated in the establishment of Aden Colony and Protectorate – now South Yemen. Even then, a high degree of local autonomy was granted to the 23 sheikhdoms in the interior.

During the 1950s and 1960s, British policy was to work towards grouping the sheikhdoms in a federation. A Federation of South Arabia was actually formed in 1959 but was opposed by the Aden Trades Union Congress and the People's Socialist Party, influenced and guided by Free Yemenis from the north. The opposition feared that the federation would eventually form the basis for an independent, but undemocratic, state.

After the republic was declared in the north, guerrilla war broke out, led by the National Liberation Front (NLF). Britain at first tried to defeat the resistance, but in 1966 announced that Aden would no longer be maintained as a naval base and that independence would be granted in two years.

In 1967, as Britain withdrew, the NLF seized power from the local rulers in the hinterland. But fighting also broke out in Aden between the

NLF and a less radical nationalist faction, the Front for the Liberation of Occupied South Yemen. In November 1967 Britain agreed to negotiate with the NLF. At the end of the month South Yemen gained independence, as the People's Democratic Republic of Yemen (PDRY), under NLF control.

There had been hopes that the two Yemens would be immediately unified, because of the close links between the republicans in the north and the nationalists in the south, but these hopes were not realised. The NLF was divided; the first president, Qahtan al-Shabi, was overthrown in 1969 by a more radical group, who brought in land reform, nationalisation and central planning. Factional differences continued to plague the NLF.

There were clashes with opponents of the new regime who, from exile in Saudi Arabia and North Yemen, tried to destabilise the country, and in 1972 and 1979 short border wars between the two Yemens broke out. However, at the end of each a commitment to eventual unity was reasserted.

Independence for the south was accompanied by a long economic crisis from which the republic began to emerge only in the 1970s. This was because independence coincided with the loss of South Yemen's main sources of income. The British subsidy to the Federation of South Arabia ended, and the income arising from Aden's role as a military base disappeared. And the Suez Canal was closed because of the Arab–Israeli war, depriving the country of revenue from bunkering and passenger services for the large passenger ships which previously used the Canal.

The PDRY developed close links with the Soviet bloc, and its relations with the West have been poor. France, Denmark and Sweden have been the only Western nations to provide bilateral economic aid.

The state took a large role in managing the economy, through successive five-year plans. The republic experienced substantial economic growth in the 1970s, mainly due to investment in infrastructure. Industrial productivity, where the state-owned sector predominates, remains low. Agriculture is based on state and collective farms. Economic aid for the PDRY has been provided mainly by China, the World Bank, the Soviet Union, and Arab states. Oil was discovered there in 1990.

Politically, merger with North Yemen remained the goal, but as recently as the mid-1980s, the obstacles to a merger seemed almost insuperable. Yet the two republics united in May 1990, partially as a result of the collapse of the Soviet Union and the discontinuance of support from Moscow.

Disputes soon broke out between the two former republics' politicians. In August 1993 a crisis was precipitated when Vice-President Ali Salem al-Biedh left the capital, Sana'a, to return to his political base in Aden. Claiming that power was being centralised in the north and complaining of mismanagement of the country's economy, he threatened to pull out of the union. Attempts to head off this prospect led to an agreement signed by al-Biedh and President Ali Abdullah Saleh in February 1994, which went some way to meeting al-Biedh's demands. But three months later civil war broke out; South Yemen declared itself independent again and sought to revert to the former status quo. Northern leaders refused to recognise the south's secession and eventually prevailed.

North Yemen

Irrigation

North Yemen is fortunate in receiving far more rainfall than many of its neighbours. But the rain does not fall uniformly, and it was soon recognised that there was considerable potential for improving agriculture by redistributing the rainwater.

Along the Red Sea coast of North Yemen is the Tihama plain, 30–60 km wide. Behind it lie highlands, whose western slopes receive the most rain and are intensively farmed. Run-off from the highlands flows through seven wadis which cross the plain. For some distance west of the highlands, there is constant flow in the wadis; further west flow occurs in irregular spates, particularly from March to October.

The Tihama plain is accessible and provides abundant land with good soil for farming and hence offered great opportunities for increasing agricultural production. In the late 1960s, institutions were set up to assist in realising this potential: the Central Planning Organisation, the Agricultural Credit Bank, and the Tihama Development Authority. Successive five-year plans set about implementing irrigation schemes on the various wadis – Wadi Zabid during the first plan (1976–80), then Wadi Rima (1980–5). The US$50 million Wadi Mawr project was developed between 1984 and 1987, following a feasibility study carried out in 1973–9.

Wadi Mawr is one of the largest drainage systems in Yemen, with a catchment area of about 8,000 sq km. It is the northernmost of the seven wadis crossing the Tihama plain, and reaches the Red Sea at Luhayyah.

The project is located on the part of the flood plain where agriculture was already supported by spate flow from the wadi – in an area of approximately 20,000 ha. Irrigation had been practised here for centuries, by means of temporary earth bunds constructed across the wadi bed to divert flow into canals, producing one of the best developed farming areas on the plain. The object of the project was to provide a permanent means of diverting flow into the traditionally built canals, to produce an assured supply allowing agricultural production to be more efficient and better planned.

Other benefits were to be provided, including a network of roads improving access between local villages and markets, and water-supplies to the larger villages both within the project area and on the fringes. Outside the project area itself, the access road system was extended to Luhayyah. The consultant for the project was Sir M. MacDonald & Partners.

A major diversion structure consisting of a single weir was built on the wadi near the point at which it emerges from the foothills. The weir spans the full width of the wadi, and comprises a concrete spillway 240 m long flanked by a 180-m earth and rockfill bank. The spillway is designed to pass a flow of 4,100 m^3/s, representing a spate flood with a return period of one in 100 years. Intake works to the canal are designed for a flow of 40 m^3/s, equivalent to 85 per cent of the mean annual flow in the wadi. Between the weir and the intake works is a scour sluice structure which enables the intake to be kept free of wadi bed material which would otherwise build up. Jung Woo Development Company of South Korea was the contractor for the diversion structure.

The diverted flow passes into the head reach canal which runs parallel to the wadi on the north side for 3.3 km, before dividing into two. From this point the north supply canal continues westwards for 19 km, while the remainder of the flow passes under the wadi in an inverted siphon before continuing west for 25 km on the south side of the wadi.

Cross regulators in the supply canals divert the flow into short lengths of primary canals which transfer the water into the existing traditional canals. There are 38 of these, one served directly from the head reach canal, 17 from the north supply canal and 20 from the south.

Drop structures control the flow in the canals. There are 70 hydraulic control structures on the system, and four major road culverts. The canal system was built by Khan Construction of Pakistan.

Project access roads, built by the China Road & Bridge Engineering Company, link the major villages with the Hodeidah–Jizan highway, which runs along the Red Sea coast and was completed in 1982. The access road system consists of a ring road around the edge of the spate irrigation area, with spurs to larger villages elsewhere. At its western end the road system connects to Luhayyah on the coast. There are 133 km of road ranging from 4.5 m to 7 m in width. Most of the roads are surfaced with gravel, except for the final 6 km to Luhayyah which uses compacted sabkha (salt flats) material, and 20 km serving the project headquarters which was given a bituminous surface dressing because of the higher traffic volumes it has to carry. Rendel Palmer & Tritton and Hunting Technical Services were associated consultants.

Water supply

Provision of potable groundwater supplies for the main villages was also included in the project. There were three problems: the limited quantity of the water; its salinity; and the risk that public health could actually be worsened by making water-related diseases more prevalent.

The approach adopted for village water supplies was therefore to provide dug wells reaching only to the upper levels of groundwater which are less saline, an approach which also limited the amount that could be abstracted. Each supply comprises a sealed well-head with a concrete-lined shaft, pumps, and water distribution points at the edges of villages. The wells were designed to be capable of being replicated by local craftsmen in other villages. Large villages have motorised pumps and elevated water storage tanks to serve multiple distribution points. In all, supplies were provided to 74 villages. Vesto of Finland was the contractor.

South Yemen

Ports

After it was colonised by Britain, Aden grew rapidly in importance as a coaling station for steamships *en route* to India, East Africa, the Far East and Australia. As steam gave way to oil power, so Aden turned over to oil bunkering. Following its policy of building refineries near the point of use, in 1951 the Anglo-Persian Oil Company (APOC, later British Petroleum) was considering building a refinery in the vicinity of Aden harbour. A reconnaissance party, which included John Palmer, a partner of Rendel Palmer & Tritton, visited the colony and identified suitable sites both for the refinery and for the associated harbour which would be needed for tankers delivering the oil. The chosen site was on Little Aden peninsula, on the opposite side of the bay from where the existing commercial harbour was situated.[1]

Rendel Palmer and Tritton was appointed consultant for the oil harbour. The oil company required the refinery to come on stream within 25 months so the harbour had to be completed in the same time. The 108-week contract, awarded to a joint venture of Wimpey and Bechtel was signed in December 1952. Rendel Palmer and Tritton had been working on detailed planning and design since the site visit in July, but at the time the contract was signed no bills of quantities existed at all.

A target cost contract was therefore adopted. It was laid down in the contract that the target price would be agreed within six months of commencement. It also provided that the contractor would earn a bonus for completion ahead of time, and would receive 25 per cent of any amount by which the actual cost of the work was below the target.

Speed was the dominant factor throughout the design and con-struction process. Close liaison was maintained between the contractor and consultant as the design progressed. Because of the difficulty of finding well-graded sand for concrete locally, steel was adopted as the material from which the jetty was to be built. Welded joints were used throughout for rapid assembly. Pile driving was carried out as far as possible from the permanent work to eliminate tide-dependent operations.

The key to the early completion of the contract was the provision of sheltered water for construction of the jetty, so the breakwater was extended into deep water as rapidly as possible. It was constructed by tipping rock, found at the cliffs at the head of the breakwater, directly into the sea from wagons; it is 1,260 m long, extends 3 m above high-water level and has a maximum height of 15.5 m.

The oil terminal has a capacity to import 5 million t annually, with four berths for 32,000 dwt tankers, two on T-headed jetties and one on each side of a finger jetty. The jetties are supported on innovative hexag-onal piles, invented because during the period of postwar steel shortages tubular piles were not available. Rendel Palmer & Tritton arranged for the South Durham Steel Company to roll a special section which could be welded together to form a hexagonal pile measuring 12 inches across the flats. This became known as the Rendhex pile and was widely used.[2]

The first tanker was berthed on 17 July 1954, and the harbour contract was completed on 10 September, three months ahead of schedule and only 22 months after the site start.

Rendel Palmer & Tritton was brought in again in 1981 to advise on rehabilitation of the jetties so that three could take tankers of up to 55,000 dwt and the fourth tankers of up to 110,000 dwt. An extra berth, also for 110,000-dwt tankers, was added. Deepening the berths and access channels required 4 million m^3 of dredging. The work was completed in 1989.

Once operational, the refinery employed 250 Europeans and 1,500 Arab and Indian workers. This made it the biggest commercial concern in the colony. A new township was built to house the refinery staff. The oil company did not want to be responsible for management of the town project, as it had been obliged to be at the Abadan refinery in Iran; the municipal authority of Aden agreed that the new town should be a colony responsibility but it had neither the funds nor the resources in its public works department to design and build a new town with all its associated infrastructure in the time available.

An imaginative solution was found. Funding was provided through a British government loan, to be paid back by the oil company; the company was appointed agent to the government with responsibility for design and construction of the municipal works (Wimpey was again appointed contractor); the colony appointed a development commissioner to direct the work, while Rendel Palmer & Tritton was appointed 'certifying engineer' with responsibility for site supervision. This arrangement displayed 'a capacity for adaptation not normally associated with the Civil Service,' said Aden's director of public works, Leonard Jackson, in a 1956 paper to an ICE conference.[3] It enabled the colony to maintain control of the development without overloading its works department. Again a target cost contract was used. Work included housing (1,000 'no-fines' concrete houses were built), public buildings, water-supply, drainage, electricity supply and a telephone network, and a new major highway to Aden.

Water supply

Extensive water-supply works were undertaken in South Yemen in the early 1980s. For Greater Aden, the French firm Sogreah had carried out master planning, feasibility studies, detail design and tender documentation. John Taylor & Sons was appointed in 1981 to carry out supervision and a design review of the project.

The works comprised a new wellfield 80 km from Aden, to produce 450 l/s, and the extension of an existing wellfield. A total of 26 boreholes was originally planned, 450 mm in diameter to a maximum depth of 150 m; this was later increased to 50 bores after additional exploratory work and hydrogeological studies.

The wellfields were connected by a pipeline 55 km long. An additional 35 km of pipes, of 200–800-mm diameter, were laid to supplement the existing distribution network in Aden. Storage was provided by four ground-level reservoirs. A generating station of capacity 1,000 kVA was built at the new wellfield. The £22 million first stage of the project was completed in 1988.

Meanwhile, in another study, Groundwater Development Consultants, a subsidiary of Sir M. MacDonald & Partners, had drawn up a plan to develop an additional supply for Aden from a new wellfield in the Wadi Tuban Delta. This led in 1984 to a commission for the parent company to review the Greater Aden master plan and extend it to the year 2010. The study involved an assessment of the condition of the existing supply system from wellfields and pumping stations to reservoirs and the distribution network.

Chloride attack on the pipes meant that considerable rehabilitation was needed. The system also required extending significantly, both because of an increase in population and shifts in population centres.

Computer optimisation was used for the design of the distribution system and also to model unaccounted-for water, and the joint use of groundwater and desalinated water.

Proposals for new works included a wellfield 30 km north of Aden to supply 10.5 million m^3/d from alluvial deposits, following an exploratory drilling programme and a geophysical survey. Other recommendations included 47 km of new trunk mains, 8 km of primary distribution, communication and control systems and refurbishment of pumping facilities. The works were later implemented although not under the supervision of the original consultant.

Elsewhere in the country, John Taylor & Sons was responsible for producing a 30-year master plan for water-supply and distribution in the Qatn-Seiyun area of the Wadi Hadhramaut. The study began in 1977. It extended over an area of 1,600 sq km, comprising 60 villages, and the main towns of Shibam and Seiyun with a combined population of 200,000. Investigations of the extent, quality and treatment requirements of potential groundwater resources were undertaken, to serve a population projected to reach 600,000 by the year 2007.

It was recommended that three main wellfields should be set up; the largest would be at Seiyun and would ultimately have 18 wells. These would be drilled between 110 m and 150 m into the cretaceous sandstone. With some additional minor wellfields the projected total demand of 47,000 m³/d (225 l/d) could be satisfied.

Storage was initially provided by elevated reservoirs, supplemented later by additional low-level storage. A trunk main system, incorporating pipes of up to 400-mm diameter, was later built to the wellfields and distribution systems. The client, the Public Water Corporation, is undertaking design of minor distribution systems itself. It is also implementing a programme of immediate works which has been in progress since the completion of the original study in 1979.

Following on from the master plan, Taylor was commissioned to undertake detailed design of a new water-supply system to the main town, Seiyun, and surrounding villages. A new wellfield with seven 150-m deep boreholes, chlorination facilities, and a main 2,700 m³ storage tank with additional 200 m³ tanks for the villages were provided. A total of 80 km of pipework was laid.

Sewerage
During the 1970s improvements to the facilities for disposing of wastewater around Aden became necessary. The Ma'alla, Tawahi and Sheikh Othman districts of Aden all discharged untreated sewage into the inner harbour at Aden, causing a pollution problem. Some parts of the Sheikh Othman district had no facilities.

John Taylor & Sons was appointed in 1975 to report on sewerage for all these areas. A final report was produced in 1977, and in 1980 and 1981 the firm was commissioned to produce detailed designs. The Ma'alla and Tawahi scheme serves a population of 70,000 and required 4 km of foul sewers of up to 700-mm diameter to be laid, together with 3 km of pumping mains. There are two pumping stations and one ejector station, and the sewage is pumped to a headworks on reclaimed land near the entrance to the inner harbour. Here it is screened, macerated and retained in holding tanks during incoming tides. It is discharged when the tide goes out, preventing pollution of the inner harbour.

For the Sheikh Othman district, with a population of 156,000, 19 km of sewers and 3 km of pumping mains were laid. Sewage flows to an existing sewage treatment works which was extended. There it is screened and macerated before passing into waste stabilisation ponds. This project was completed in 1986.

Makulla is the second largest town in southern Yemen and is the administrative capital of the Hadhramaut Governorate. Here, in addition to a sewerage scheme, land reclamation and coastal protection works were also needed. The town is built on a narrow strip of land between the sea and the mountains, so that properties on the edge of the sea were prone to damage from waves, especially during the monsoon season. No land was available to construct interceptor sewers.

The solution was to build sea walls with reclaimed areas behind them, simultaneously protecting existing properties and providing space for interceptor sewers, pumping stations, ejector stations and a future road. Sewage is pumped to a headworks designed to cope with a population of 75,000, and is discharged into a long sea outfall 450 mm in diameter and 1,200 m long. A total of 2.5 km of coastal protection works was constructed and 213,260 m³ of landfill was used in the land reclamation. Construction took place between 1985 and 1989, with Danish contractors, partly funded by the Danish government and partly by international aid agencies.

In Seiyun, the population of 30,000 was expected to double in the 25 years from 1986. Domestic wastewater was disposed of to soak pits which were becoming overloaded because improved water-supply had led to increased consumption. A network of sewers was planned which would deliver sewage to stabilisation ponds, from which the final effluent would be suitable for use in agriculture for irrigation. As city-wide sewerage proved too expensive to implement at once, community septic tanks were planned with a municipal emptying service.

Roads

Rendel Palmer & Tritton was engaged in building roads in both North and South Yemen. A major project in the North was the World Bank-funded upgrading of the Sana'a–Hodeidah highway, which crossed spectacular escarpments with heights varying from sea level to 3,350 m. Rendel Palmer & Tritton was responsible for all stages from design to tender assessment; it lost the commission for supervision after the World Bank withdrew its loan following a disagreement with the Yemeni highway authority. The funding was replaced from Arab sources and the supervision contract given to a different consultant.

However, Rendel Palmer & Tritton did supervise the 62 km Marib to Safer road which provides access to the newly established oilfields, as well as a road between Marib and Harib. 'It was primarily a military road and went all round the mountains,' says Rendel Palmer & Tritton assistant director Gordon Matthews. 'Part way through construction the two countries united, so there was little need for it.'

In the south the consultant was appointed by Sun Oil of Texas to design and supervise construction of access roads for its oil concession in the Shabwa area. A total of 200 km of gravel-surfaced roads was built, capable of taking vehicles of up to 60 t, carrying well-drilling equipment, to wherever the oil company decided to drill. This involved going 'up wadis and escarpments, with S-bends and hairpin bends,' says Matthews.

A complete walk-over survey of a 2,600-sq km area was carried out by Rendel Palmer & Tritton engineers using Magella transmitters to fix their position relative to three satellites. Aerial photos were also used to help determine the easiest routes. Detailed surveys of the most promising areas followed. Geometry of the roads was determined by the ability of the oil company's vehicles to negotiate bends. The roads were surfaced with a 150-mm wearing course of graded and rolled gravel.

Yemen today

The 1994 civil war in Yemen raised questions about its future. Though the North eventually prevailed preventing the South from seceding, it failed to achieve the quick victory in the war which many had expected. This seemed to support the view that the south, perhaps as a result of colonisation and later Soviet involvement, has achieved a greater level of organisation and technical capability than the north, which is still highly tribal. This in turn suggests that attempts to hold the unified republic together may not succeed in the long term.

chapter eleven

SUDAN

―――

Sudan, the largest country in Africa, is also perhaps one of the least fortunate. For all but 10 years since independence in the mid-1950s it has been riven by civil war between the Arab-dominated north – latterly with a hard-line Islamic government – and the African ethnic and largely Christian groups of the south. A period of stability and apparent prosperity in the 1970s left the country with unserviceable levels of debt and effectively bankrupt; drought and severe famine have dogged the country since the 1980s.

Sudan's economy is mainly agricultural, with cotton being the most important export. Oil in limited quantities has been discovered in south and central Sudan but the civil war has prevented its development, and Chevron, the oil company involved, has pulled out.

Sudan is regarded as part of the Middle East because 60 per cent of its population is Arab, its political destiny has always been closely linked with that of Egypt, and the official religion is Islam, though only 70 per cent of the population is Muslim.

Sudan encompasses a wide range of ethnic groups – the country has only been considered a single entity since early in the nineteenth century, and for a quarter of this century there were plans to separate the south from the rest of the country. Historically, Sudan's main attraction to outside powers was as a source of gold and slaves.

In 1820 the ruler of Egypt, Muhammad Ali, conquered most of the north and central regions of Sudan in the name of the Ottoman empire, and Turkish–Egyptian rule lasted until the rebellion in 1881 led by Muhammad Ahmad ibn Abdullah, known as the Mahdi. This did not suit British interests, because of the strategic importance of the upper waters of the Nile – whoever controlled these had a disproportionate influence over Egypt, which at the time was effectively a British colony. After a long campaign, General Herbert Kitchener's Egyptian army, enhanced by 8,000 British troops, defeated the Mahdists at Omdurman in 1898. (Kitchener's Island, opposite the Cataract Hotel at Aswan, is today a poignant reminder of the campaign.)

Sudan then came under joint Anglo-Egyptian rule, with Britain being the dominant ruling partner, until it gained independence in 1956. Between World Wars I and II a policy of preparing the south for eventual membership of a federation of East African states was pursued, and links between north and south were discouraged for 25 years. By the time this policy was reversed, it had had a lasting and damaging effect on the prospects for national unity.

After World War II, Sudan's future became a major source of disagreement between Britain and Egypt. Egypt claimed Sudan for itself and sought to unite the two countries under Egyptian rule, while Britain wanted full independence. The situation remained deadlocked until Nasser came to power in Egypt and dropped claims to Sudan – hoping, nonetheless, that, once independent, Sudan would choose to be united with Egypt. Meanwhile, the British hoped that a pro-British regime would be elected. In fact both were disappointed. Elections brought victory for the pro-Egyptians, but the government decided on full independence rather than unity with Egypt. After a transitional period starting in 1954, Sudan became fully independent on 1 January 1956.

Meanwhile, in 1955, a mutiny in a southern garrison had triggered civil war, caused by disaffection in the south arising from discrimination against southerners in the army and civil service. This war raged for 17 years, with only one serious attempt by one of various governments to end it by dialogue rather than force. In May 1969 a bloodless coup by junior army officers brought Colonel Jaafar Nimeiri to power.

Nimeiri pursued socialist-oriented policies including nationalisation, which led to the withdrawal of expatriate businesses and people. But in 1972 he achieved a political master-stroke, ending the civil war through a peace formula which offered the south a form of regional self-government.

This left Sudan's future looking bright. A period of optimism began which lasted for most of the 1970s and gave Nimeiri the opportunity to concentrate on the economy. He planned to turn Sudan into a major exporter of food to the Arab world, combining Arab finance and western technology with Sudan's natural resources. The principle seemed sound – Sudan has 80 million ha of potentially cultivable land, of which only a small fraction is actually farmed.

But the plan proved over-ambitious and went horribly wrong. Large numbers of peasants and nomads were displaced to make way for large-scale farming and irrigation schemes, which were dogged by soil erosion, silting of canals and growth of weeds. Sudan was unable to produce cotton of competitive quality and its exports declined. There were attempts to switch to wheat, for which conditions in Sudan were not really suitable. The country found itself saddled with crippling debts: even Saudi Arabia and Kuwait had provided cash in the form of loans rather than grants. Sudan suffered from a chronic lack of foreign exchange and had difficulty paying for oil imported from Iraq.

Development was also hampered by shortcomings in Sudan's basic infrastructure, so money had to be found not just for new projects themselves but for infrastructure improvements as well. In 1978 *New Civil Engineer* reported that there was only 400 km of metalled road in the country, though the almost completed Port Sudan–Khartoum highway would treble that. Port Sudan had reached full capacity, and the railway system, though good by African standards, needed an estimated £900 million to be spent on it to bring it up to modern standards.[1]

In the early 1980s, as Sudan lapsed into effective bankruptcy, Nimeiri gradually lost his grip on the country through a series of unpopular decisions. These included the introduction, in 1983, of Islamic law, which was strictly enforced, and the removal of food subsidies in 1985 at the insistence of the International Monetary Fund. Civil war broke out again in the south, initially because the regional government had proved unworkable and corrupt. This was fanned into a significant uprising by the unpopularity in the south of the imposition of Islamic law.

Nimeiri was deposed while on a visit to Washington DC in the US, and was replaced with a transitional military council until elections could be held in 1986. The government lacked the resources to relieve the drought and famine which then struck and for which international agencies had to shoulder responsibility. The economy remained paralysed by debt; even after rescheduling, and committing half the total government expenditure for 1985/6 to servicing interest payments, no significant impression was made on the debt burden.

A fundamentalist Islamic government came to power in 1989. Little or none of the debt, now standing at US$16,000 million, has been paid back.

Roseires dam

Sudan is primarily an agricultural country. Rainfall over much of the country is sufficient only for subsistence agriculture; development of the economy, which required the ability to produce crops for export, has depended on extensive use of irrigation.

The confluence of the Blue and White Niles is at Khartoum. The Blue Nile is the source of the Nile's annual flood, which provides most of the water for irrigation in both Sudan and Egypt. But 66 per cent of the Blue Nile's flow occurs during August and September, and while the peak flow is on average 6,300 m^3/s, in the low season between January and April it can be as low as 100 m^3/s. Since crops need to be brought to maturity during this period, storage reservoirs are essential to provide irrigation water.

Sudan's first large irrigation project was the Gezira Scheme, designed by Sir Murdoch MacDonald himself, and inaugurated in 1925 when the Sennar dam on the Blue Nile was completed. Initially water from Sennar was used to irrigate an area of 300,000 feddans (acres) between the Blue and White Niles; this was successful and the irrigated area expanded until by 1952 it reached a million feddans. It was further extended by the Managil scheme between 1956 and 1962, which brought a further 800,000 feddans to the south and west of the Gezira scheme under cultivation. But it had been recognised that after completion of the Managil extension, further water storage capacity would be essential.

As far back as 1904, the building of a dam at Roseires, south of Sennar on the Blue Nile, had been suggested by Sir William Garstin. In 1955 it was decided that the time had arrived for a serious study. Accordingly Sir Alexander Gibb & Partners was commissioned to investigate a proposal to construct a dam capable of storing 1,000 million m^3 of water near Roseires.

The consultant was also instructed to report on how much land was suitable for irrigation on the Blue Nile between Sennar and the border with Ethiopia, along the White Nile between Kosti and the Sobat river and in the area of the Rahad and Dinder rivers. After two years of study Gibb produced a preliminary design for a dam of the required capacity, capable of being heightened at a later date if necessary. It identified 1.75 million feddans of irrigable land, of which 1.25 million feddans were in the Kenana area.

It was clear that Roseires was urgently needed. In 1955, Gibb and Coyne & Bellier of Paris were jointly appointed consultants to design a much bigger dam than that originally planned. This was to have a capacity of 3,000 million m^3, capable of being increased later to 7,400 million by raising the height of the dam by 10 m. This was almost ten times the original storage capacity of Sennar.

Development of Roseires depended on a new Nile Waters

Agreement being concluded between Sudan and Egypt. An accord reached in November 1959 defined the share of the total annual flow to which each country was entitled. It allowed Egypt to build the Aswan High Dam, while Sudan could build Roseires and any other works needed to use its share of the water effectively.

The total irrigable area which Sennar and Roseires (stage I) combined could serve, based on an 80 per cent flow year in the Blue Nile, was calculated at 3.5 million feddans, the increase due to Roseires being 1.3 million feddans. With the additional stage II heightening of Roseires, under the 1959 Nile Waters Agreement the total irrigable area would rise to 4.3 million feddans. In the absence of restrictions a potential 6.1 million feddans could be irrigated.

Since there must always be a flow through the dam to support downstream projects, including Sennar, there was also potential to develop hydroelectric power. A hydroelectric power station with seven 30-MW turbines was designed, to be built as a separate project from the dam and capable of supplying 960 million kWh from stage I. Merz and McLellan was associated consultant for the electrical and mechanical installations.

A loan worth £S18 million (£18.45 million) for construction of Roseires (stage I) was provided by the World Bank's soft-lending arm, the International Development Association (IDA), and a West German bank.

Even while this loan was being negotiated, preliminary work had started on site, financed directly by the Sudanese government. This included the construction of a township for 2,000 single and 552 married workers, permanent houses for staff of the government departments involved, plus a hospital, offices, canteen and other ancillary buildings.

A permanent all-weather airfield, a network of roads, a generating station and a water treatment plant were built by Motherwell Bridge of the UK and Contracting & Trading Company of Beirut. The 'black cotton soil', a highly plastic clay, overlies the site and softens considerably when wet. 'It became impassable during the rainy season. It was an essential early part of the works to get down roads so we could move around the site,' says John Bowcock, the recent chairman of Gibb. Supplies were brought in from Khartoum by rail.

Construction of a prestressed concrete road-and-rail bridge, with reinforced concrete approach viaduct, across the Blue Nile downstream of the dam was also started at the same time. It was built by a consortium of Christiani & Neilsen of Denmark, Skanska of Sweden, Grun & Bilfinger of West Germany, and Kier of the UK. Problems with aggregate delayed completion of the bridge until the second season of the main dam works in 1962–3 (see below).

The main dam contract was awarded in 1961 to the Italian consortium Impregilo. The dam consists of a central concrete buttress section around 1 km long, flanked by very long earth embankments, 8.5 km on the west and 4 km on the other side. Mass concrete gravity sections form the transition between the buttresses and the embankment sections. The buttresses are founded in trenches up to 18 m deep excavated through weathered rock to reach solid rock below. Foundations for the possible future heightening of the dam (stage II) were built at least up to tailwater level as part of stage I.

One of the problems for dams on the main Nile or Blue Nile is the large amount of silt washed down the river in the annual flood. Roseires has five deep sluices in the main river channel, each with radial gate openings measuring 10.5 m x 6.0 m. These are used to regulate the flow through the dam under normal conditions, and because they are positioned at a low level they allow most of the silt carried with the current to pass through the dam. To the west of these is a surface spillway with seven 13-m x 10-m radial gates. These come into operation at the peak of the Nile flood, and because they are offset from the main sluices the flow is spread over the width of the river bed much as it was before the dam was built. Floating debris brought down the river by the flood passes over the spillway and downstream. A stilling basin was carved out of the natural rock 60 m downstream. A maximum design flood of 18,750 m^3/s can be passed without the dam being overtopped, compared with the maximum recorded flood in the 60 years before construction of 10,800 m^3/s.[2]

In a typical year water is impounded in the dam during the tail end of the flood, in October, raising the reservoir level from 467 m to 480 m above datum. This stores almost 2,400 million m^3 of water which is drawn down between November and the following May.

A small hydroelectric station is positioned between two buttresses adjacent to the spillway structure, to provide power for operation of the gates and for the township at Damazin. This is separate from the main hydropower station, built after the main dam between 1967 and 1971, to supply Khartoum.

Headworks were provided at each end of the dam to supply water to later irrigation projects. On the west bank a 360 m^3/s canal was to be built to supply water to the Kenana project; to the east an equivalent headworks was built to supply the Dinder and Roseires project areas. In the event neither side has been developed and the headworks remain closed off by concrete bulkheads.

Materials for constructing the embankment dam sections came from local sources. Borrow pits provided fill; the top layer of black cotton soil was often suitable for the impermeable core of the dam. Lower sand deposits were used in the dam shells and berms. Embankment slopes were protected by gravel quarried from deposits on the river banks, with dumped riprap for additional protection, for which rock came partly from the excavations for the diversion channel and partly from quarries.

Work on site was restricted by the Nile flood and the working season extended from October, when the river level had fallen enough for work to begin, until July when the site would be inundated. Impregilo programmed to complete the work in five seasons.

In the first season a natural flood channel was enlarged to form a diversion channel. 'We selected the site because of the channel – it enabled the river to be pushed about,' says Gibb's Bowcock. An island in the channel divides it in two, and here it was possible to start work on the spillway section in dry conditions during the first season. Elsewhere, aggregate and concrete plants were set up and abutments for temporary bridges needed for tipping in the next season were completed. Work on the permanent bridge continued.

In the following season the temporary bridges were built on their abutments to allow two rockfill cofferdams to be built in the main channel. Once the space between the cofferdams had been pumped out work could begin on the deep sluice section. Work also began on the west earth embankment, and completion of the permanent bridge gave access to allow work to begin on the east bank. The cofferdams were flat in profile and were protected with large rocks and gabions and a torpedo net to allow them to be safely overtopped during the second season's flood.

In the third and fourth seasons the earth embankments were brought to their finished level, while the concrete section of the dam was brought to crest level except for two gaps either side of the spillway section to allow the floodwaters through. With the sluices finished, at the end of the fourth season the rockfill cofferdams could be removed. Next season cofferdams were built in the diversion channel to allow the remaining gaps in the buttress section of the dam to be filled. The spillway stilling basin was excavated and the foundations for the power station constructed. On 1 July 1966 the sluices were closed and the dam began to impound water. The final overall cost of Stage I of Roseires was £24.8 million.

Roseires bridge

The Roseires bridge over the Blue Nile formed part of the Roseires dam project. The bridge is situated downstream of the dam at a height of 9.5 m above maximum water level, allowing traffic to cross at any time of the year. Although initially intended for road traffic, it was designed so that a single narrow gauge (1.06 m) railway track could be added in the future. It was intended to do this in such a way that vehicles could still use the bridge between trains.

Consultants Gibb and Coyne & Bellier had originally sought tenders for a structural steel bridge. The winning tender, however, was for an alternative prestressed concrete design by Christiani and Neilson and Skanska, submitted by a joint contracting venture in which the designers were joined by Grun & Bilfinger and Kier as contractors.

Three main spans of 45.42 m, 90 m and 45.42 m are flanked by two approach spans to the east and a 13-span viaduct to the west. The approach spans are made up of three prefabricated T-section girders side by side, weighing 80 t each. These were cast and post-tensioned longitudinally in a yard on the west bank and erected by two self-propelling gantry cranes and a hydraulic launching girder which was moved from pier to pier as construction progressed. The box girder main spans were constructed as free cantilevers, and were post-tensioned both longitudinally and transversely.

As completed in March 1963, the bridge has a hinged joint at the centre of the 90 m main span. For conversion to a railway bridge, ducts were provided to allow additional prestressing cables to be added, and the hinged joint at central span was to be fixed. The planned conversion has not, so far, been implemented.

Kenana irrigation scheme

The Kenana sugar plantation scheme was one of Sudan's priority projects. Costing £300 million, it was intended to turn sugar cane, a major import, into an export crop. The original idea is attributed to Tiny Rowland of Lonrho, who commissioned a feasibility study.

Kenana is about 240 km south of Khartoum on a plateau of Blue

Nile clay. Originally the scheme was to have been irrigated from the Blue Nile, but because of depleted supplies this proved impossible. Nevertheless it was shown to be economically feasible to pump water up 45 m from the White Nile instead.

Consultant Howard Humphreys & Partners, now part of Brown & Root, became involved in 1973 in a pilot scheme and later designed the complete canal network. It had to win over the Sudanese authorities to new ways of designing canals. Under the specifications of Sudan's irrigation department, to which all schemes had hitherto conformed, canals were straight and with shallow embankments. Howard Humphreys proposed smaller section canals, often on high embankments. They were designed to follow contours, using computers to give the shortest distance to suit the lie of the land and to optimise cut and fill. 'The authorities found it difficult to accept that canals could be "contoured" or that their embankments could be built as high as we wanted them to be,' Howard Humphreys' chief resident engineer John Tolbutt told *New Civil Engineer* in 1978.[3]

Eventually, however, the principles were accepted and Howard Humphreys was commissioned to design irrigation and ancillary works for a scheme to produce 3.5 million t of sugar from an estate of 33,000 ha. This included a 29-km main canal from the White Nile to the estate, 279 km of primary distribution canals, over 500 take-off structures to regulate the flow of water, and four pumping stations. The primary canals are the ones which employ the principle of contouring; rough designs initiated at Howard Humphreys' Reading office and prepared using a Sudanese 2-km square grid survey were checked and adjusted on site, and returned to Reading for final computer optimisation.

The UK contractor Sir Alfred McAlpine was awarded a contract in 1975 for 19 km of main canal, 30 km of access roads, an airstrip, and the Kenana township with services for a population of 7,000. For the major muck-shift involved in building the embankments it imported 50 items of heavy plant which were brought 1,000 km across the desert in a convoy. 'The locals had never seen so much traffic,' McAlpine's project mananger Andrew Hollway told *New Civil Engineer*, 'nor so many desirable batteries or starter motors. Security was a real problem . . . we lost whole vehicles, not just components.'

In June 1977 McAlpine was awarded a second contract, for the civil works of the sugar refinery, for a lump sum of US$20 million. Its original canal contract was extended to include the first batch of primary canals, and was renegotiated to be placed on a lump-sum rather than a cost-plus basis. The value was US$51.5 million. McAlpine was disappointed not to win the work for the rest of the canals but the policy of the Sudanese authorities was to award work to publicly owned companies, to save foreign exchange, once it became clear that the work was within the capabilities of the local firms.

Most of the canal embankments were formed of the black cotton soil found on the site. 'Black cotton soil fissures badly when it is dry, but it is highly expansive and self-sealing when it is wet,' said Hollway. Only the main canal stilling basins upstream of the pumping stations, which were founded on embankments of gravel, had to be lined in concrete.

Commissioning of the first stretches of main canal, irrigating an area of 2,000 ha, began in 1977, only 16 months after the contractors' plant arrived on site.

Rahad irrigation

At around the same time as Kenana, the even more ambitious Rahad scheme was being developed. Rahad was Sudan's largest irrigation project, covering 125,000 ha. Sir M. MacDonald & Partners was appointed consultant on the scheme. The firm's involvement in Sudan dates back to the 1920s and the early stages of the Gezira project, on which Sir Murdoch MacDonald himself worked. Other, later, partners in the firm began their careers working on irrigation projects in Sudan.

Rahad, which cost £196 million at 1978 prices, was a completely new irrigation and drainage system with associated infrastructure, intended to produce cotton and groundnuts over an area of 126,000 ha. It lies on the east bank of the River Rahad, which flows seasonally between July and November. A permanent water supply is obtained by pumping from the Blue Nile at Meina. From there water is carried 85 km by canal to a barrage and head regulator on the Rahad. *En route* to the Rahad, the supply canal passes under the River Dinder in a reinforced concrete inverted siphon. The main irrigation canal is 90 km long and supplies water to around 5,700 km of minor canals.

The project also included the provision of associated infrastructure for new townships, warehouses and workshops, with 160 km of roads, electricity transmission, and telecommunications. Work began in 1977 and associated consultants on the project were Hunting Technical Services, Rendel Palmer & Tritton and Mott, Hay & Anderson.

The road network connects the townships throughout the area with the main Khartoum to Port Sudan highway and consists of 80 km of primary road and an equal length of slightly narrower feeder roads. Soil found on the Rahad plain is expansive clay, and the roads are built on embankments made of quarried fill. Structures on the network include a number of pipe culverts where the roads cross irrigation canals, and 10 highway bridges. A five-span reinforced concrete bridge carries the primary road over the River Rahad.

In the 1970s cotton production in Sudan declined. This was a serious matter given that the crop accounted for at least 50 per cent of the country's exports. Sudan found it difficult to produce, economically, cotton of quality good enough to compete with that grown by the US, Thailand and China. In the early 1980s Sudan's Ministry of Finance & Economic Planning commissioned a major marketing study, involving Sir M. MacDonald & Partners, Mott, Hay & Anderson, the British Cotton Growing Association, accountants Peat Marwick Mitchell and Company, Hunting Technical Services and a local consultant.

The government wanted to increase production to 2 million bales of lint by 1991. The study looked at how the processing and marketing side of the operation, from collection of seed cotton in the field to sale and shipment, also needed upgrading to cope with production at this level. A detailed inventory was made of existing facilities: 39 ginning factories; storage, testing and other operations. Pest control was also investigated.

A report was produced in 1984, giving financial and economic assessments of capacity and quality improvements. Spending of £44 million over six years on ginning facilities was proposed. Some of the recommendations were put into effect – for example, Sir M. MacDonald supervised the refurbishment of six cotton ginning factories at a cost of £7.3 million, completed in 1989 – but in general implementation has been piecemeal.

Roseires dam heightening

Roseires was designed with in-built provision for raising its crest level by 10 m at a future date, without the hydroelectric plant or hydraulic equipment needing to be modified. Several studies have investigated whether heightening was economically feasible at any given time. In 1980 a consortium of Gibb, Coyne & Bellier, Hunting Technical Services and Sir M. MacDonald & Partners (as coordinator) produced the Blue Nile Waters Study. This considered the whole question of the availability of water and land in the Blue Nile region. Initially it aimed to identify a package of developments which together would use the available irrigation and hydropower potential to the full. It then went on to look at the consequences of heightening Roseires by up to the 10-m maximum. The findings were incorporated into the Nile Waters master plan for the whole river system, but no immediate action to heighten the dam followed.

In 1985, Gibb, in association with Merz and McLellan, looked at the question again and studied three options. These were: raising the dam by the full 10 m (an option which could be commissioned, it estimated at the time, by 1994); raising the dam by an intermediate amount, with the possibility of earlier commissioning; and raising by 10 m in two stages, the second stage being commissioned by 2005.

Gibb looked at the benefits arising both from the increased irrigation which heightening would make possible, and from the additional hydropower which would save burning fuel in thermal power stations and allow construction of new thermal generating plant to be deferred. These benefits were projected to 1995. Gibb also used its Nile System computer model to project the state of irrigation development and of reservoir siltation to 2005.

The analysis suggested that heightening by only 7 m produced optimum gains. Compared with raising the dam by the full 10 m, this required less investment, gave a higher return, could come into operation a year earlier, and reduced evaporative losses. There was no advantage in heightening to any lower figure, and making provision to raise the dam by the remaining 3 m as a separate stage was not thought to be economically justified.

However, the client also took other, non-economic factors into account and in 1992 gave instructions to proceed with heightening by the full 10 m. Gibb again worked on the design with Coyne & Bellier.

Raising the dam requires the addition of a new 10 m high concrete section on the crest of the existing concrete buttress section, in addition to which the buttress webs are to be extended and thickened. The earth embankments on each side will be almost doubled in length to 23 km.

Reservoir siltation is being modelled, and the development of the as yet unused canal headworks for power and irrigation is being looked at again. A key factor in design and construction of the heightened dam is the requirement to keep the reservoir fully operational throughout. Tender documents were being produced in late 1993 after which detailed design was set to commence, with the intention of having the higher dam ready to impound the 1998 flood.

Water and sewerage

Thanks to its location at the confluence of the White and Blue Niles, Khartoum has always had a plentiful supply of fresh water, which has been piped to most areas of the city.

Sewers and a sewage treatment works designed by Howard Humphreys, but serving only part of the city, were constructed by Marples Ridgway from 1960, and sewers and primary treatment works serving other small areas were constructed later. Proposals for a modern city-wide scheme were being canvassed in the later 1980s, but financial and political difficulties have prevented their implementation.

Sudan today

Sudan's future does not look hopeful at present. The government is universally unpopular, both internally with Muslims and non-Muslims alike, and externally. Almost no development aid is reaching the country: western governments are concerned about the debt problem, the current regime's human rights record, and the civil war; while Arab states, notably Saudi Arabia, see the Islamic government as a threat. Lack of hard currency is a problem. There seems little prospect of an immediate change in Sudan's fortunes.

LIBYA

Libya is the fourth largest country in Africa, but most of its area is desert. It has a relatively small population and until the discovery of oil it was one of the poorest countries in the world. Oil wealth brought profound changes to the lives of the Libyan people, but revenue collapsed in the 1980s and limited income has been a constraint on the country's development since then. Libya depends heavily on imports, especially food imports. For this reason the most important recent and ongoing project is the Great Man-Made River, which transports water for irrigation from aquifers in southern Libya to the coastal zone. Increased agricultural output resulting from this scheme will help to reduce dependence on food imports.

Libya's origins extend as far back as recorded history; over the centuries it was settled by Phoenicians, Greeks, Romans – Libya (which was once fertile) has been described as a 'Roman granary' – and by Byzantium. Between the seventh and twelfth centuries AD it was gradually converted to Islam under the influence of the Arab occupation; during this time Tripoli port became a terminus for Saharan trade routes.

The country came under nominal Ottoman control in the early sixteenth century, but enjoyed considerable autonomy. This ended in 1911 when Italy occupied Libya and, after facing considerable opposition, eventually began colonisation of the fertile areas of Jabal al-Akhdar and the Jafara plain. Italy was forced out in World War II, and a joint British–French administration was installed which continued after the war. The three provinces of Cyrenaica, Tripolitania and the Fezzan were administered separately and each developed its own sense of identity.

The United Nations called for independence to be granted by 1952 and the chosen system was a federal monarchy under King Idris al-Sanusi. This favoured British interests as the new king was sympathetic to Britain, and allowed, for example, a UK and US military presence to remain. The rise of Nasserism in neighbouring Egypt, however, encouraged opposition to the Libyan regime which came to a head in 1967 when the government failed to make a stand over the Arab–Israeli war.

A coup by junior officers two years later brought Muammar Gadaffi to power, leading a Nasserist and Arab nationalist government.

The new regime removed power from the country's traditional ruling élite groups. In pursuit of Arab unity it also made several attempts at political union with other Arab states. These included a federation with Egypt and Syria in 1972 and a short-lived merger with Egypt in 1973. British and US troops were ordered out of the country, oil prices forced up, and diplomatic relations with the Soviet Union established. The regime survived attempted coups in 1975 and 1984; despite the introduction of nominally democratic political structures, Gadaffi and his collaborators remain in effective control.

Libya has aroused hostility in Europe and the US by its support for the Palestinian cause. Economic sanctions were imposed after Libya refused US and British demands to hand over the alleged perpetrators of the bombing of the Pan Am aircraft over Lockerbie in Scotland in 1988. Within the Middle East the Libyan government is also often unpopular because of its unpredictable stance on regional and international issues.

Oil exports began in 1961, bringing wealth for the first time to what was a poor country. Libyan crude oil is light and highly marketable. Nevertheless, revenues collapsed in the early 1980s, from US$21,000 million in 1980 to US$4,000 million in 1986. This led to a cutback in spending and a concentration on large single projects such as the Ras Lanuf refinery and petrochemicals complex, the Misurata steel plant and the Great Man-Made River.

There are state-imposed limits on personal wealth. Most Libyans have seen their standard of living rise since oil exports began; there has also been a radical change from a predominantly rural society to one concentrated in urban areas, particularly around Tripoli and Benghazi.

Since the Gadaffi regime came to power, many British firms have seen Libya as commercially risky and have found it imprudent to seek work there.

Water supply and irrigation: the Great Man-Made River

The Great Man-Made River is claimed to be the world's biggest civil engineering project. It is certainly of utmost importance to Libya.

The fertile areas of the Jafara Plain, the Jabal al-Akhdar and the coastal plain of Sirte are separated by a band of semi-desert from the Sahara, which steadily and ominously advances a few more metres every year. Water supplies in the fertile strip along the coast are under increasing pressure. There is no reliable surface water: the 350 mm of annual rainfall is not readily captured and used. Farmers therefore have to rely on groundwater. In 1949 abstraction and recharge were in equilibrium but since then abstraction has grown until it is now four times the volume of recharge. Levels in the aquifer have been progressively drawn down, requiring more pumping and adding to the cost of irrigation, while quality has also suffered and there has been saline intrusion in places. With the country already heavily dependent on imports of food, it was obvious that this lack of water would become an increasingly serious problem.

Ironically, the solution lay in the desert. During oil exploration in the 1960s, drilling rigs found an equally valuable resource: immense underground water reserves. Studies were instigated which confirmed the extent of these aquifers, proving the existence of four vast basins of good quality water. The question then became how best to exploit this resource.

One solution would have been to set up irrigation schemes in the desert. However, considering that the rural population was already leaving the land and migrating to the cities in large numbers, it was unlikely that many people would be keen to begin farming in the desert. Instead, the possibility of bringing the water to the centres of population along the Mediterranean coast was investigated. This would be an immense undertaking, requiring water to be piped hundreds of kilometres. But studies showed that even with the massive investment required to build the pipelines, this solution was considerably cheaper than desalination of sea water.

It was decided to go ahead with the vast investment needed, and in October 1983 the Great Man-Made River Authority was set up. The multinational engineering, project management and construction company Brown & Root was appointed engineering and project management consultant to the authority and implementation of phase I of the project followed immediately.

The Great Man-Made River is remarkable not so much for its technological sophistication as for its sheer scale. Five phases are planned the second of which is in progress and the first has already been completed. It is estimated that the total cost of all phases will be US$25,000 million.

Phase I delivers 2 million m^3/d of water to the coastal strip between Benghazi and Sirte. Phase II will supply 1 million m^3/d of water to the Jafara plain. The other three phases are supplementary, expanding the project to cover all areas of the country. Phase III will increase the flow in the phase I system to 3.68 million m^3/d. Phase IV will supply water to Tobruk along a pipeline from Ajdabiya. Phase V will link phases I and II via a pipeline from Sirte to Tripoli, and an additional two wellfields will increase the water flow in phase II to 2 million m^3/d.[1]

The four underground basins are, first, the Kufra basin in the south-east, which covers an area of 350,000 sq km and contains an estimated 20,000 km^3 of water. It connects with the Sirte basin which holds an estimated 10,000 km^3. The Murzuk basin extends from Qargaf Arch in the north of Libya to beyond its south-western borders. Its total area is 450,000 sq km, storing 4,800 km^3. Finally there is the Hamadah and Jufrah basin which extends from the Qargaf Arch to the coast. The basins were formed up to 38,000 years ago at a time when Africa's climate was temperate and there was heavy rainfall.

Water quality in the Murzuk and Kufra basins is very good, with a total dissolved solids content of 300 ppm in the Murzuk basin and 250 ppm in that of Kufra. Quality in the Sirte and Hamadah basins reduces as it approaches the coast, where sandstones give way to marine limestones which can have a high salt content.

Wellfields to extract the water are being built 400–700 km inland both to tap the better quality water and because the aquifers are nearer the surface there. The wellfields are designed to extract water at a rate which will not excessively lower the level in the aquifer, keeping pumping costs to a minimum and allowing the wells to remain productive longer. The initial rate of extraction of 2 million m^3/d of water from the eastern wellfields and 1 million m^3/d from the western fields provides several hundred years' worth of potential production. A small amount of recharging of the aquifers occurs through the heavy but intermittent rains that fall on the southern uplands. The difference in level between the wellfields and the coast means that flow is largely by gravity.

Phase I is the largest component of the project. A total of 1,840 km of prestressed concrete cylinder pipe carries water to the coast of this total 1,530 km are 4 m diameter. The pipeline begins at a wellfield in Tazerbo where there are 108 production wells, of which 98 are sufficient to produce the flow rate of 1 million m^3/d, the others remaining on standby. They are connected at 1.3-km spacings to three parallel collector pipelines, 0.6 to 1.6 m in diameter and 10 km apart. These feed into a larger spine collector which connects to a 170,000-m^3 header tank at Tazerbo.

A second wellfield of 126 production wells is located 256 km north at Sarir. The wells are arranged in a similar way to those at the

Tazerbo field and provide an additional 1 million m³/d of water. Wellfield modelling was used to determine the number of wells, their depth and water-level and the quantities of water which could be removed without jeopardising production. At Tazerbo and Sarir the wells are about 450 m deep, with submersible pumps at a depth of 145 m. Control units at each well-head receive signals from the central communication and control system.

Collector pipelines lead to two header tanks and thence to the main Sarir–Sirte pipeline. This consists of two parallel 4-m diameter pipes through which water flows by gravity to the Ajdabiya holding reservoir, 380 km to the north. The holding reservoir has a capacity of 4 million m³ and consists of a circular earth embankment 9 m high and 930 m in diameter. From here two pipelines run west and north, to Sirte and Benghazi respectively. The Sirte pipeline can deliver 820,000 m³/d by gravity which can be increased to 2.3 million m³/d by pumping. The Benghazi pipeline has a capacity of 1.18 million m³/d by gravity flow, and a maximum of 2.5 million m³/d under pumping. At the end of each pipeline is another circular earth-embankment balancing reservoir with capacities of 6.8 million m³ at Sirte and 4.7 million m³ at Benghazi.

Operation of the system is on the principle of continuous supply at constant rates throughout the year, so large strategic storage reservoirs have also been built to store surplus water during winter for use during summer when demand is higher. There are two of these reservoirs, with a capacity of 37 million m³ in the Sirte area and 76 million m³ at Benghazi.

Prestressed concrete cylinder pipes were found to be most cost-effective for constructing the pipeline, with most of the raw materials for them being available in Libya itself. They consist of a steel cylinder embedded in a concrete core, wrapped with steel wire and coated with mortar. Additional external protection is provided where there is a high concentration of chlorides or sulphates in the ground, and cathodic protection will be provided if it proves necessary after a few years of operation.

The British consultant Sir Alexander Gibb & Partners became involved in the project when it was approached by the US firm Price Brothers to check design of pipelines and structures for phase I. Price Brothers (UK) was acting as sub-contractor for the pipe manufacturing plant to the South Korean Dong Ah consortium which had won a design-and-construct contract for 1,840 km of pipeline in phase I. Subsequently Dong Ah approached Gibb directly to design the 930 m diameter Ajdabiya holding reservoir. When Dong Ah subsequently won work on phase II, Gibb was brought in as consultant from the start.

The pipes are made at purpose-built factories at Sarir and Brega. Each section is 7.5 m long and weighs up to 86 t. A total of 236,000 pipe sections were used in Phase I.

A permanent communication and control scheme will direct operation of the entire project from Benghazi, issuing instructions to five local subsidiary operation, support and maintenance centres from which the actual hardware will be remotely controlled. A central computer system at Benghazi will monitor operation of the system.

Flowing at an average speed of 0.95 m/s, water will take over nine days to travel from Sarir to Sirte or from Tazerbo to Benghazi. The system's response to changes in flow is slow and any changes need to be planned well in advance, so effective monitoring is essential. For the initial few years of operation of the project, the system is to be operated manually, with computer assistance in decision-making.

Power for phase I is provided by a 90-MW plant at Sarir which consists of six gas turbine generating sets. Phase II is currently under construction, and is due for completion in 1998. It will deliver 2 million m^3 of water from wellfields in the Jabal Hasouna region to the western coastal belt, particularly the Jafara plain. Water will be delivered from three pumping stations to terminal reservoirs at Tarhunah and Garabulli. Draw-offs will be provided along the pipelines to supply water into wadis for agriculture.

Phase II will be controlled by computer from a control centre at Ben Ghashir.

Although the Great Man-Made River project will provide water for domestic, industrial and municipal users, it is planned that most, 86 per cent, will be used in agricultural production. Crops such as wheat, barley, sorghum and sheep fodder are to be given priority. There will be a mixture of smallholdings, of a size designed to support a single family, and large farms and cooperatives of 1,600–2,000 ha. When complete, the project will allow irrigation and development of 155,000 ha.

Industrial development, especially the large schemes of Ras Lanuf and Brega, is expected to be encouraged by the availability of water at much lower cost than it would have been from desalination. The shortage of good-quality water which affects most Libyan cities will also be alleviated.

Mott MacDonald is working on the main contract for phase II as consultant to the Daewoo Corporation of South Korea. The design package includes civil and mechanical and electrical work, hydraulics, surge analysis, operation and maintenance and training. The scheme encompasses manufacturing plants for the prestressed pipes, living accommodation, water and power supply and roads. A computer simulation model of the scheme which can be used in training as well as for design purposes has also been developed. It features a graphics-based user-interface which allows the program to be used by people without previous computer experience.

The firm is also working on a master plan for utilisation of water from phase II of the project in the Tripoli region, looking in particular at how water from the project can best be used in conjunction with local groundwater sources which need to be allowed to recover from over-abstraction. 'The aquifer will take an enormous time to recharge,' says Mott MacDonald's hydraulics engineering divisional director John Pavey. 'Even using Man-Made River water alone would not allow it to recover quickly.'

Ports

Several British firms have been involved in the extensive port development which has taken place in Libya. In the early 1960s, Sir William Halcrow & Partners was engaged in reconstruction of the port at Benghazi, which had been seriously damaged in World War II. Damage had subsequently been exacerbated by the action of the sea. Unexploded bombs had to be removed from the harbour before work could begin.

Work on phase I started in 1961. This included repair of the breakwaters (with complete reconstruction of the outer one); building a new 320-m deep water quay for general cargo; dredging, reclamation, and the provision of a new oil berth. Halcrow's newly developed Stabit concrete armour units were used for the first time on this contract, to protect the breakwaters. The 29-t units successfully withstood 10-m high waves in winter storms during 1962–3.

Under phase II, begun in 1964, two more deepwater quays totalling 445 m in length were added. The total cost for phases I and II, including dredging and provision of transit sheds, was £6.6 million. Later work included design of port administration buildings.

Rendel Palmer & Tritton was later responsible for extending the port further. Breakwaters totalling 4 km were built in water up to 17 m deep. Some 120 ha of land was reclaimed within the port and

3 million m^3 of material was dredged. Another 3 km of blockwork quays, plus facilities for container, roll-on roll off services and oil products traffic were also provided.

The port of Tripoli had also been severely damaged during the war and though rehabilitation was urgently needed, the Libyan government at the time was severely short of funds. Halcrow was able to come up with a scheme to make the port usable again for only £613,000. For this, tanker berths were provided in addition to basic reconstruction work.

In 1970 Sir Bruce White, Wolfe Barry & Partners was commissioned to carry out a feasibility study for the port's further development. Commissions for detailed design and supervision followed.

Stage 1B, as it was designated, consisted of 2 km of rubble-mound breakwater and 2.6 km of concrete caisson quays. Extensive land reclamation was carried out within the breakwaters, requiring 4 million m^3 of reclamation material, partly provided by extensive dredging works. Two slipways of 500-t and 1,000-t capacity were also provided, plus warehouses, a cold store and ancillary buildings. The £50 million contract was carried out between 1973 and 1979.

Following the production of a master plan in 1974 the firm went on to design and supervise stage 2. This involved building an outer harbour formed by 4 km of rubble-mound breakwaters enclosing an area of 400 ha. Within this area are 4 km of reinforced concrete caisson quays and 1 km of rubble bunds. A terminal for refined oil products capable of accommodating tankers of up to 40,000 dwt was provided as well as warehouses, workshops, passenger accommodation and administration buildings. A total of 9 km of roads was also built. Work was let in a number of separate contracts. Of these, contract 2A (for the breakwaters) was valued at £45 million, and that for dredging the new approach channel at £15 million.

A proposal for a corniche ring road to improve access to the port was taken to tender documentation stage before being postponed.

Rendel Palmer & Tritton was responsible for the design of the first phase of the new port of Ras Lanuf. This is sited at the end of the Gulf of Sirte and serves the major new industrial complex including an oil refinery, petrochemicals plant and an associated township.

The deepwater harbour is protected by 5 km of rubble-mound breakwaters armed with 48-t tetrapods – Rendel's answer to the Stabit. Phase I included construction of the breakwaters, the longest of which is 2.5 km, and of six berths for the export of liquid products, connected to the shore by a 3-km piled pipe track and roadway. A 600-m quay provides facilities for containers and general cargo. It was constructed of 21 concrete caissons, each weighing 5,500 t, cast on dry land and floated into position.

A second phase included the provision of additional berths for crude oil export and an extension of the quay wall to allow export of other products from the industrial complex. Mott MacDonald was brought in as consultant by the client during phase II.

Mott MacDonald is currently involved in the reconstruction of the breakwater at Tripoli, which was damaged by storms during the 1980s. It took several years for the Libyan authorities to reach a decision on repair or reconstruction of the breakwater. Mott, Hay & Anderson, as it then was, was called in by the Yugoslav contractor which had been appointed to carry out the work, to redesign proposals originally produced some time earlier by a Dutch consultant.

Roads

After Colonel Gaddafi came to power in 1969 the cities of Tripoli and Benghazi expanded rapidly as new commercial and industrial areas were built, as well as new houses, schools and hospitals. This, with the country's increasing prosperity, led to a large growth in car ownership and in demand for travel, placing severe strain on the cities' transport infrastructure. This was exacerbated by the construction of new residential areas far from places of work which meant that workers had to travel long distances. Traditionally, homes and places of work had been near each other.

Rendel Palmer & Tritton was commissioned to plan and design 50 km of urban motorway in Tripoli, and supervise construction of around half that length comprising the second and third ring roads. The work included 20 grade-separated interchanges, 70 bridges, 12 footbridges, lighting and signs. Rendel Palmer & Tritton was also responsible for the initial site survey. Detailed design took only 12 months, though there were some subsequent modifications because of changes in the master plan for the city.

Design work on the Tripoli project started in 1976 and construction began in 1980. The third ring road has not been completed because of lack of funds. Rendel Palmer & Tritton's assistant director Gordon Matthews, who was responsible for supervision from 1980 until the opening of the system in 1988, says that one of the biggest problems, in the initial stages at least, was the standard of asphalting and concreting. 'In hot countries like Libya, when concrete has to be transported very far, it's difficult to get it in place before the initial set which occurs in 90 minutes or less. Retarders were used, but if too much was added the set could be delayed until the next hot cycle the next day,' he says. If the concrete gets too hot it dries out and fails to hydrate properly. The results are immediately apparent in the form of cracking on the surface of the concrete, which reduces its durability. 'We had quite a lot of problems,' says Matthews, 'but the contractors were very cooperative.' On early defective work the problem was remedied by coating the concrete with a crack-sealing epoxy. 'Subsequent work was improved by changes to the mix, better curing, preventing the concrete from drying by erecting tents around it, pouring at night and so on.'

In Benghazi, Rendel Palmer & Tritton was responsible for design and supervision of construction of the third ring road, including seven grade-separated interchanges and 12 bridges. The firm also designed three major grade-separated junctions, the al-Uruba interchanges, of which only two were completed.

The overall roads strategy within the Benghazi master plan was the responsibility of Ove Arup & Partners. This firm was asked in late 1977 to submit proposals for a traffic study of the city followed by design and supervision for 50 km of road. After the contract began, this modest length expanded somewhat dramatically to 700 km, later cut back to 600 km. As a result, Ove Arup was 'effectively acting as the traffic and highway department of the Municipality of Benghazi', the company's director Malcolm Simpson said in *Arup Journal* in 1980.[2]

Just as engineers for water and sewerage schemes found that their first task was to produce some estimate of the population to be served, Ove Arup first needed to undertake a traffic study and make an assessment of likely travel demand. The framework was an overall land-use master plan for the city produced by the US planners, Whiting Associates. The existing road network consisted of paved major roads and unpaved minor roads, all suffering from a severe lack of maintenance. Among the objectives of Ove Arup's traffic study were the following: to produce an efficient private and public transport network for all users; to provide a transport infrastructure flexible enough to allow for alterations in the city plan; to enhance safety, especially of pedestrians; and to retain and enhance interesting facets of the existing urban townscape.

Simple extrapolation from existing traffic movements was not adequate; changes in land use, population and social habits also had to be accounted for. Development of the plan therefore included an inventory of existing roads; establishment of journey speeds; counts of vehicles by category; roadside interviews; and a parking survey.

Origin and destination studies revealed that only a small percentage of journeys began or ended outside the city, demonstrating that congestion problems in the city would not be solved by a bypass, as some analysts had argued. The population was found to be rising faster than the master plan assumed, and Ove Arup predicted that this, with increasing affluence, would produce a steep increase in demand for vehicles which, if unrestrained, 'could not possibly be accommodated within the current context of city plans and policies.'

After studying four options – unrestrained and restrained car use, in combination with central or decentralised employment – Ove Arup concluded that a policy of car-use restraint with decentralised employment would reduce city centre congestion and utilise the proposed ring-road system most efficiently. This analysis was accepted by the city authorities, allowing for the first time a roads programme with a logical basis to replace the *ad hoc* approach which had previously existed.

The first contracts were let in 1980. They were won by nationalised Libyan contractors and international firms, mainly Italian, Greek, Yugoslav and Korean. As the master plan covers the period to 2000, work is expected to continue for some time.

Another important part of the Benghazi town plan was the Corniche project, intended to improve the inner harbour by defining its shoreline; the project also included a new 4.5 km dual carriageway corniche expressway. The corniche was formed by building a new sea wall and dredging the inner harbour to a greater depth. The dredged material was used to reclaim land for the road and to create recreational open spaces near the harbour.

A key feature of the expressway is the high-level Guiliana Bridge over the inner harbour which consists of two single cell prestressed concrete box girders of variable depth side by side, one for each carriageway, with three continuous spans of 80 m, 120 m and 80 m. These are flanked on either side by two 35-m viaduct spans. Halcrow designed the scheme, which was built between 1968 and 1975 at a cost of £5.2 million.

Buildings

Ove Arup was involved for many years in the construction of the various phases of al-Fatah University, Tripoli. This was based on an existing establishment and was originally planned as a college of advanced technology. Ove Arup became involved in 1968 on what was to have been a teacher-training college, but during design this was renamed the Faculty of Education. Architects James Cubbitt and Fello Atkinson & Partners produced a master plan for a university, originally known as the University of Libya, Tripoli but whose name was later changed to al-Fatah University. Ove Arup was responsible for structural design and services, the latter being initially sublet.[3]

The university, which was planned to take 10,000 students, occupies a vast 246 ha site approximately 3 km east of Tripoli. Most facets of education are catered for with around 20 individual faculties for the arts, science, engineering, education, agriculture, and so on. In addition there are central facilities such as libraries, administration buildings, dormitories and a large central power plant.

Phase I comprised the faculties of education and agriculture, dining halls, and a later addition, the high voltage laboratory. The education faculty alone comprised 13 buildings which were generally two-storey. Ove Arup concluded that use of reinforced concrete was the only viable means of construction because of the expense and difficulty of obtaining imported steel sections, and the lack of local expertise in prestressed concrete. In general, framed construction was used with some buidings having additional shear walls to provide lateral stability. Repetition in the designs, especially for the faculty of agriculture's laboratories, encouraged precasting.

Most striking structurally and architecturally is the faculty of agriculture's administration building, a two-storey block built on a circular arc with a circular dome over the main conference room, supported off an elliptical slab, and with asymmetrical column supports.

Tripoli is in a low-activity seismic region. In the absence of local regulations, Ove Arup designed the buildings to Californian codes of practice, with advice from Professor Ambraseys of Imperial College, University of London. Ove Arup later incorporated standard details into an earthquake design manual so that they could be used in future projects in earthquake zones.

Sewerage

John Taylor & Sons undertook sewerage schemes in several Libyan towns. In 1975 the firm was asked to produce proposals for a treatment works at Sebha, the administrative centre of the Fezzan region. This was to replace the existing, overloaded works and provision was to be made for using treated effluent in irrigation.

Phase I of this scheme served a population of 35,000 while phase II was designed for a projected population of 80,000. Tertiary treatment was by rapid gravity sand filters and chlorination. A pumping station conveyed treated effluent to a storage lagoon 2.5 km away.

At Yeffren, 110 km from Tripoli, there were no existing facilities, so a separate system for foul and surface water was proposed. The foul-sewer system includes two pumping stations and 11 km of sewers, connecting to an extended aeration treatment plant. The treatment works was built in 1980–5 in two phases, each designed to cater for 11,500 people.

At El-Igelat, 80 km south-west of Tripoli, there were no sewers but there was a treatment works 8 km away at Sabratha. John Taylor & Sons designed a separate foul and surface water system, with an intake works for Sabratha. The foul system incorporates five pumping stations, 34 km of sewers and 3 km of pumping main. The surface water system comprises 14 km of sewers discharging to soakaways.

Libya today

In recent years the Gadaffi regime has moderated its policies both internally and externally. The result has been improved relations with European trading partners, especially Italy (which needs oil and gas from Libya), France and Germany. Many believe that this apparent change is merely opportunistic. In the Gulf crisis Libya appeared to waver for several days over whether to support Iraq's invasion of Kuwait before eventually denouncing it.

Relations with Britain and the US remain strained, however, especially over Libya's supposed involvement in the Lockerbie bombing. In these circumstances, UK firms seeking work in the country are likely to remain the exception.

> Expatriates living and working in the Middle East were no doubt able, for the most part, to put out of their minds the volatile political nature of most of the places in which they found themselves working and to concentrate on the job in hand. But *coups d'état* were a way of life in the region and expats could, and did, find themselves in potentially perilous situations.
>
> Viv Hoad of Halcrow worked in Libya for 12 years in the

1950s and 1960s. In 1969 he was nearing the end of a two-year secondment as executive director of the Beida Development Organisation, in charge of a £50 million programme of government buildings. Hoad and his family were provided with a villa in Benghazi, while Hoad also had a flat at Beida itself, 200 km away. He recalls:

'One morning – 1 September 1969 – I set out from the villa in Benghazi to get to the office for 8 a.m. . . . . I left at 5.30–6.00. I got to Beida to be met by barricades across the road. I was told there'd been a *coup*, and I couldn't go to my office – the army had taken over. I set off back in my government car to my family. Just before Benghazi I was stopped by revolutionaries coming out in their armoured cars who asked me what I was doing. By then there was a curfew in force. Eventually I persuaded them to allow me to continue. . . .

At the end of the highway I was stopped by a young soldier. He said: 'No, you can't go on.' He jumped in the car and I had to drive him round various parts of Benghazi for half an hour, commandeered. Eventually I persuaded him to let me leave the car at a petrol station and walk. There were various water trenches open and I cowered in these when I heard any shots . . . A man called Widgery, the director of electricity for the whole country, lived on the ring road, next door to Halcrow's chief resident engineer who was away during the *coup*. I reached Widgery's house and found several other expats there, and my son. I spent the night there. Next morning the curfew was lifted. My villa was half-a-mile away. Somebody gave me a lift . . . to my relief, my wife and daughter appeared while I was there . . .'

Fortunately the coup was fairly bloodless. After about 10 days, government officials were ordered to report to their workplace. Hoad retrieved his car and set off for Beida, negotiating roadblocks *en route*, where the soldiers were 'very courteous', and reported to the police there. 'Next day I was stopped again on the way to the office. It was all sealed and padlocked. I met the old director of administration there who said: "Go back to the flat and wait." There was no way of contacting my wife. . . .' Luckily Halcrow's deputy chief resident engineer found Hoad there, and provided some company and reassurance.

'Several days elapsed. Then on the Friday afternoon there was a knock at the door – I was told: "You are going to be expelled from the country tomorrow." Hoad returned to Benghazi and went to the emigration office and consulate-general to make arrangements for leaving. 'Then I was told I couldn't leave, because the authorities needed to get information from me about the project. After a few days I was summoned to Beida to be investigated.' His family was allowed to leave without him, however.

After another anxious period of waiting Hoad was eventually interviewed by revolutionary officials about the detail of some of the contractual payments under Beida contracts. 'They asked questions such as why had I paid for some setting out which had not been done (it had, but the pegs had been stolen by Bedouin), and why had I not deducted liquidated damages from one of the contractors for being late . . . Afterwards I returned to Benghazi, where I had to wait until my agreement ran out on 4 December – I was not allowed to leave until then: Gadaffi didn't want to be construed as having sacked anybody.'

c h a p t e r   t h i r t e e n

EGYPT

_____

The Valley of the Nile and the Nile Delta in Egypt have for centuries been highly fertile regions. It was there that some of the world's earliest civilisations arose, 8,000 years ago.

Because the engineers and architects of ancient Egypt used stone, more of their magnificent constructions are still standing than those of other early civilisations. While the pyramids of Giza and the temples of Karnak are the best known, there are many other equally impressive and remarkable structures.

One that is unusual because it displays a high order of modern, as well as ancient, engineering skill is the Great Temple of Rameses II at Abu Simbel in Nubia. Constructed inside a mountain about 1,250 BC and over 60 m long, it created awe in the local populace because, on the Pharoah's birthday and again on his wedding anniversary, but on no other day, the rising sun shone a beam of light on to the faces of the statues at the furthermost end of the temple.

This was possible because these dates were the same number of days on either side of the winter solstice, and the astronomer-priests calculated the angle and direction of the rising sun on those dates and had a hole drilled in exactly that direction through the eastern-facing wall at the entrance to the temple.

The construction of the Aswan High Dam in the 1960s would have flooded the temple, and Unesco called for proposals to protect it. The scheme adopted was designed by the Swedish consultant VBB and involved hewing out the temple walls from the mountain and re-erecting them in an artificial cliff-face above the high-water level of the new reservoir. Because of the higher level of the temple in its new position, the dates on which the sun strikes the faces are now out by one day. Visitors to the site, after viewing the temple, are taken behind the scenes to see the modern engineering marvel – the artificial cliff – and are as impressed with it as with the work of the ancient engineers.

Another modern discovery which highlights the variety and scale of the works constructed in the days of the Pharoahs is the 'Boat beneath the Pyramid'. Excavated in 1954, this was the Royal Barge of King

Cheops, for whom the Great Pyramid was built. The barge is of graceful design, 43 m long and 6 m wide, and was used to convey the dead Pharoah's body along the Nile from Memphis, the then capital of Egypt, to the Valley Temple on the west bank of the Nile. This was joined by a causeway to the pyramid which was to be his final resting-place.

Among other ancient engineering works were the Nileometers which measured the height of the Nile at various locations along its course, and which the Pharoahs used to determine the rate of tax which farmers had to pay each year.

Despite Egypt's good fortune in having the fertile lands of the Nile Valley and Delta, however, cultivable land represents only 5 per cent of the country's total area. And despite a history of attempts at industrialisation, dating back to the early nineteenth century, it has remained a relatively poor country with a large peasant population.

Egypt has modest oil reserves and production reached a peak of 910,000 b/d in 1985, but reserves are estimated sufficient to last only a few years at current production rates. Most of Egypt's income is derived from oil exports, the Suez Canal tolls, tourism receipts, and remittances from emigrant workers, though it has also received substantial aid from the US and, to a lesser extent, Europe.

Egypt became a Muslim country in the seventh century AD. From the time of Alexander the Great in the fourth century BC until 1952 it came under a succession of foreign rulers: the Greek, Roman, Byzantine, Arab and Ottoman empires. From 1805 until 1849 it was ruled by Muhammad Ali, a Macedonian, ostensibly on behalf of the Ottoman empire but, in reality, as an almost independent state.

Muhammad Ali was the first ruler to try to industrialise the country. His attempts were closely connected to his military ambitions in the region which were thwarted by Istanbul with British help when he appeared to be becoming too powerful. However, his dynasty remained in power and his grandson Ismail continued the process of modernisation which Muhammad Ali had begun. Unfortunately this process saddled Egypt with unmanageable debts. One result was that Ismail was obliged to sell his 44 per cent holding of Suez Canal shares to Britain in 1875. This brought only a temporary respite and Egypt was declared bankrupt in 1876. The result was Anglo-French, and later simply British, economic control from then until 1936.

In 1882 Britain invaded Egypt to suppress resistance to the Khedive (the ruler), and although France prevented Britain from formally including Egypt in the British empire it became, effectively, a British dependency.

Egypt gained limited independence following a unilateral British declaration of 1922 and a treaty in 1936, although British interests were protected and British troops remained stationed there.

By the end of World War II calls for British withdrawal were growing, with King Farouk, monarch since 1936, increasingly seen as ineffective. Despite the withdrawal of British troops to the Suez Canal zone, opposition, and guerrilla activity, escalated. Eventually the Free Officers' movement, led by Gamal Abdel-Nasser, took over in a coup in 1952 and exiled the king. An agreement was eventually reached in March 1956 under which British troops were withdrawn.

Nasser opposed closer links with the West – he was a co-founder of the Nonaligned Movement – and relations with western powers deteriorated. In 1955 an arms deal with Czechoslovakia was announced after attempts to buy weapons from the US, Britain and France failed because they had been made conditional on unacceptable political concessions. In July 1956 the US persuaded the World Bank to withdraw an offer of funding for the Aswan High Dam. In retaliation Nasser nationalised the Suez Canal Company, prompting the Suez crisis and the invasion of the Canal Zone by Israel, Britain and France. Compensation for the Canal owners was later agreed. But for almost the next two decades, Egypt came under the influence of the Soviet Union.

Thus it was the Soviet Union which supplied funding and technical support for the Aswan High Dam. In the 1960s a new drive towards industrialisation began, with the nationalisation of a number of key private companies. This process was again underpinned by large amounts of Soviet aid although balance of payments deficits limited progress.

Nasser died in 1970 and was succeeded by Anwar Sadat. A shift in alignment towards the West occurred in conjunction with Sadat's policy of negotiating the return of Sinai, occupied by Israel since the Six-Day War in 1967 and which Egypt had attempted to regain by force in 1973. Sadat accepted US mediation, because of Washington's influence with Israel.

In 1974 Sadat introduced the 'open-door' policy designed to encourage foreign investment. A new statute, Law 43, allowed joint venture companies to be set up, provided the local partner held at least 51 per cent. This provided substantial opportunities for British firms to work in the country. In 1975, the second of two disengagement agreements with Israel allowed the Suez Canal, closed since 1967, to reopen. The Sinai oilfields were handed back to Egypt.

The open-door policy did not benefit Egypt's economy as much as had been hoped. It attracted mainly oil companies and banks; extensive state control of other areas of the economy discouraged outside investors. The policy also caused a rise in imports.

Higher oil revenues and aid from the US kept Egypt's balance of trade in surplus in the early 1980s. Arab aid was cut off after the 1978 Camp David accords with Israel and Egypt remained largely isolated from the rest of the Arab world. In the mid-1980s, when oil revenue fell, help from the International Monetary Fund was needed.

British civil engineering involvement in Egypt dates back to the middle of the last century when Muhammad Ali's immediate successor, Abbas, allowed James Stephenson to build the Cairo–Alexandria railway in 1850–1. However, Abbas was an exception, in general Muhammad Ali's dynasty was Francophile and it was a Frenchman, Ferdinand de Lesseps, who was responsible for the Suez Canal. A palace built for the Empress Eugénie of France when she visited Egypt for the formal opening of the canal was converted in modern times into the Marriott hotel. The old palace was modernised to become the public rooms while two tower blocks were added to provide the bedrooms.

In 1890 the British engineer William (later Sir William) Willcocks, who spent a large part of his life in Egypt and Iraq and was at the time working for the Egyptian government's irrigation service, was instructed to survey the Nile and propose a site for a dam to regulate it. Willcocks identified the site where the first Aswan dam was built early in the following century. He also established the principle for providing the dam with sufficient sluices to allow the peak of the Nile flood with its load of sediment to pass through.

An international commission consisting of Sir Benjamin Baker, designer of the Forth Railway Bridge in Scotland, the French engineer Auguste Boulé and the Italian Giacomo Torricelli endorsed most of Willcock's proposals. Sir Benjamin was later appointed consulting engineer to the Egyptian government to oversee its construction. Willcocks was a man of strong opinions and his highly readable autobiography, *Sixty Years in the East*, contains an amusingly sardonic account of the friction between the personalities on the international commission, especially between himself and Boulé.[1]

He later clashed with Baker over the raising of the dam, Willcocks being in favour of this and Baker casting doubt on its safety. Eventually Willcocks' view prevailed; the dam was in fact heightened twice, in 1912 and 1933.

In 1902 the resident engineer on the Aswan dam was Murdoch MacDonald, who went on to become director-general of reservoirs with Egypt's ministry of works, overseeing the heightening of the dam. He was responsible for many irrigation and drainage projects in Lower Egypt, and later for the Gezira irrigation system and the Sennar dam in Sudan. In 1922 he founded the consulting firm of MacDonald and MacCorquodale which later became Sir M. MacDonald and Partners, and which continued to work extensively in the Middle East.[2]

Lacking large oil resources, Egypt has depended heavily on aid since World War II. This came variously from the West, the Soviet Union, and other Arab states. As a result of this dependence Egypt did not enjoy the huge boom in construction that characterised the rest of the Middle East in the 1970s and early 1980s. Nevertheless several major projects were carried out, including the Greater Cairo Wastewater scheme. This became the biggest urban sewerage scheme in the world, as well as the biggest British overseas project of the mid to late 1980s.

The Aswan High Dam

In the 1950s, the British consultant Sir Alexander Gibb & Partners was initially commissioned to design the Aswan High Dam. Kennedy & Donkin was also engaged on designing the hydroelectric power side of the scheme. The US and Britain had agreed to provide US$270 million to finance the first stage of its construction but western involvement ended after the World Bank cancelled its offer to contribute in 1956.

Nasser's initial justification for nationalising the Suez Canal was that five years' tolls would pay for the dam. In fact it was built with Soviet aid at a total cost of US$1,000 million. The rockfill dam is 111 m high and 3,830 m along its crest. Lake Nasser, formed by the 165,000 million m^3 of impounded water, is 400 km long.

In addition to impounding the annual floodwaters of the Nile allowing them to be released on irrigated land when most useful, the dam improves navigation on the lower Nile and generates 2,100 MW of hydroelectric power.

The dam, opened in 1971, has received considerable criticism. Particular causes of controversy have been the high evaporative losses from the lake, up to 10,000 million m^3 annually, and the fact that the dam impounds the 40 million t of silt carried down each year by the flood. Previously every year the Nile floodwaters had inundated agricultural land in the valley, depositing the silt which contributed to the valley's fertility. The loss of this silt was in itself a cause of criticism, and it has also been widely suggested that siltation would limit the life of the dam. The original dam was equipped with 180 sluices to let the peak flow, and with it the silt, pass through.

In fact, a conference of the International Commission on Large Dams held in Cairo in 1993[3] was told that only about 1 km³ of silt has been deposited in Lake Nasser since 1973, out of a total capacity of 168 km³. Evaporative losses were anticipated; they are far outweighed by the benefits of security of supply and total control of the annual flood, whose effects were not wholly beneficial. Agricultural production has been increased by 20 per cent and the dam saved Egypt from famine during droughts in 1972–3 and 1979–87. And 143,000 million kWh of electrical energy was generated by the Aswan hydro-electric power station between 1967 and 1991. Professor Tony Allan of London University's School of Oriental & African Studies has said that the economic impact of the dam has been 'resoundingly positive' while some of the environmental impacts for which the dam has most been criticised 'cannot be shown to be unambiguously negative'.[4]

Reconstruction of the Suez isthmus

British firms had to wait for Sadat's 1974 open-door policy, and improved relations between Egypt and Israel, before they could find significant work opportunities in Egypt. The next major priority became the clearing and reopening of the Suez Canal, which had been the dividing line between Israeli and Egyptian troops and was littered with mines and sunken ships. Also on the agenda were improvements to the infrastructure, particularly for water and wastewater, of the main cities, Cairo and Alexandria.

A high-level Advisory Committee for Reconstruction had been formed to coordinate work on the country's infrastructure under the auspices of the new Ministry for Housing & Reconstruction. The plan was to create a new axis of population and development across the Suez isthmus, with 'free zones' for industry and export at Port Said and Suez. It was planned to double the cross-sectional area of the canal in a two-stage widening scheme to make it passable by the new supertankers. The plans were costed in 1974 at £3,600 million.

Because Egypt had little recent history of working with western organisations, the committee appointed the US consultant Tippetts-Abbott-McCarthy-Stratton (TAMS) as overall adviser. A small team from TAMS, working with the committee, had the job of putting together the terms of reference for commissions, letting the work to consultants and eventually getting construction in hand.

Shortly after the committee was formed, partners from the UK consultant Bullen and Partners (now Bullen Consultants) made a speculative trip to Egypt. Terms of reference for the reconstruction of Port Said, at the northern end of the Canal, had just been issued. The project called for new harbour work, which was Bullen's speciality, but it also called for economics, town-planning skills, and general rebuilding of infrastructure.

On returning to the UK, Bullen assembled a team which it subsequently led and which included Binnie & Partners on the infrastructure side, Peat Marwick Mitchell (now KPMG) as economist, and Shankland & Cox as town planner. The team's bid won the commission.

The team worked as an integrated unit to consider various problems and requirements. These included: a population rise to 750,000 by the year 2000, and the consequent need to provide employment for it; the development of a new port to handle 10 million t/y of cargo by the same date; provision of a transport system linking the port with the national system; the need for adequate provision for development of industry, tourism, recreational and social services; and the rehabilitation of existing engineering services, especially water-supply and waste disposal, and their future augmentation. The team was required to present a status report after six months, an evaluation after nine, a master plan after a year, and a final revised plan after 15 months. During the commission, the terms of reference were extended to study a free zone at Port Said.

David Kell, now a partner with Binnie, spent 1975 in Egypt working on the project. 'Port Said was a ghost town,' he recalls. Its population had shrunk from 200,000 to 20,000. Not only the Canal but also the beaches were mined. 'My job was to look at what needed to be done immediately to get the water-supply and wastewater systems back into a condition to be able to serve the population as it returned.' Electricity and solid waste also came within Binnie's brief.

Similar commissions for the towns of Ismailia and Suez were also won by British consortia. The Suez team was led by Halcrow with planning consultant Robert Matthew Johnson Marshall & Partners and Economic Consultants. Architect and planning consultant Clifford Culpin & Partners led the Ismailia team, with Louis Berger International of the US as engineering consultant and Oficina Tecnica de Empresas e Ingeniera of Spain advising on land reclamation. Oxford University's Economic Institute was economic adviser, and Transportation and Environmental Studies advised on environ-mental matters.

As part of the reconstruction of the Suez isthmus a £50 million road tunnel under the Canal was built 12 km north of Port Suez. The project comprised a 1.63-km long concrete-lined bored tunnel of 11.6 m diameter to carry a two-lane road. Named after the Egyptian military engineer Ahmed Hamdi, its depth was 50 m below water-level to allow for the plans for deepening and widening the Canal. Halcrow was consultant, and the contractor was a joint venture of Tarmac Overseas of the UK and Arab Contractors. As the first permanent crossing of the Canal, it was intended to allow exploitation and development of the east bank's mineral wealth.

Greater Cairo Wastewater Project

During 1975 terms of reference for major wastewater projects in Cairo, Alexandria and Helwan became available simultaneously. On the Port Said project Binnie & Partners had worked with Professor William Hanna of Cairo University as local associate. Binnie's David Kell asked him to suggest a specialist to act in a similar capacity for wastewater. He proposed Dr Abdul Warith. On visiting him Kell discovered, firstly, that another UK consultant, John Taylor & Sons, had separately been recommended to Dr Warith; and, secondly, that consultants had been invited to attend a site visit and briefings in Cairo and Alexandria prior to submitting bids for the wastewater schemes.

Representatives of 30 firms from all over the world were taken by coach to the two cities. 'During these trips everyone was sizing up the opposition,' says Kell. During the briefing sessions the Taylor and Binnie representatives separately decided that it could be advantageous to join with another firm to add strength to the bid. 'We got together over a beer and decided to form a joint venture, Taylor Binnie & Partners,' Kell says. On the Cairo scheme and certain other projects the two firms have worked together ever since.

The two firms' joint venture put in proposals for all three cities. 'We were told we were rated first on Cairo and Alexandria, but we couldn't be awarded both,' says Kell. The firms decided to go for the biggest scheme, the Cairo project. (The Alexandria project went to Camp, Dresser & McKee of the US, whose scheme included a proposal to discharge sewage into the Mediterranean via a long sea outfall. This met with opposition from the Egyptian government and academics at Alexandria University. Instead a sewage works was built, with the aim of using the effluent in agriculture.)

Cairo's original sewage system had been designed by the British engineer J. Carkeet James and built between 1910 and 1915. The scheme served the east bank of the Nile, where most of the population then lived. It was designed for a population expected to reach about one million over the following 25 years. It was enlarged in the 1920s by the British engineer Albert Pinson, and again in the 1950s; but by the mid 1970s it was having to cope with a population of nearly eight million people.

The original network comprised 39 km of property connections and 270 km of street sewers feeding to a main collector 1.6 m in diameter. Because of the low-lying nature of the ground, over 64 pneumatic ejector stations were needed to feed the sewage into three sealed mains which fed the collector. Flow then gravitated 13.6 km to Ein Shams, from where it was pumped a further 12 km to a primitive sewage farm at Jabal el-Asfar, north-east of Cairo on the edge of the desert. Here it was filtered before, mostly untreated, effluent was discharged into the Mediterranean via Lake Manzala.

In 1924–9 a second, and larger, main collector was added for a length of 5.3 km to a new major pumping station at Ameria, with a new 19.2-km pumping main to Jabal el-Asfar. The first sewers to serve the west bank of the Nile, where the population was rapidly expanding, were built at this time. In 1935, Godfrey Taylor, a partner of John Taylor & Sons, advised that a proposed third collector was needed as a matter of urgency. World War II intervened and this was not built until 1950, by which time the population had grown to 3.35 million and only 45 per cent of the total area of the city was provided with sewers.

Despite a crash programme of improvements in the 1960s and the construction of an activated-sludge plant in 1970, by the time Taylor Binnie had completed its investigation, the fact that the basic system had been designed for a much smaller population could not be disguised. Sewage regularly overflowed into the streets of Cairo at over 100 locations; by the 1980s sewage flow was approaching 1.25 million m^3/d, of which only 220,000 m^3/d received secondary treatment. Partly treated effluent was discharged into the Nile Delta or the irrigation drains that crossed the Delta on their way to the Mediterranean, where significant populations drew their water supplies. Infant mortality was high: a UNICEF study in 1986 estimated that 14 per cent of children died before the age of five, about 10 times the rate of infant mortality in western countries.

The population was expected to increase to some 20 million by the end of the century; and water usage, and thus the volume of wastewater produced, was growing as Cairo's water-supply mains were extended and living standards improved. Overloading of the sewers was made worse by the accumulation of settled sand in the inverts.

Serious corrosion had affected the main collector, because of the long retention time of sewage in it, and the high ambient temperature. In the absence of oxygen, septic conditions developed, producing hydrogen sulphide which dissolved in water to form sulphuric acid. The collector was 'believed to be in parlous condition in parts' said a 1985 Institute of Civil Engineers Proceedings paper.[5]

It seemed only a matter of time before a major water-borne epidemic hit the city. In the same paper, Dr Donald Mackay of the Ross Institute said: 'The underlying aim of the whole project is not simply "to get rid of sewage". It is to dispose of a highly (and potentially explosive, in epidemiological terms) dangerous material in such a manner as to minimise the danger to the public in every possible way.'

Taylor Binnie & Partners' master planning for the Greater Cairo Wastewater scheme began in 1977. It was to develop into the largest public health improvement scheme in the world. The initial plan was finished in early 1978 and accepted by the client, the General Organisation for Sewerage & Sanitary Drainage (GOSSD), which evolved into the Cairo Wastewater Organisation (CWO). The client wanted to make a start on design and construction immediately, but after the Camp David accords Egypt had become isolated and the Kuwait-based Arab Fund for Economic and Social Development, which had funded the Taylor-Binnie study, was not prepared to fund the implementation of the scheme as had originally been envisaged. The project came to a grinding halt.

Later in 1978, the British and US governments, through the UK's Overseas Development Administration (ODA) and the US Agency for International Development (USAID), came to an agreement with the Egyptian government to co-fund the project. Britain promised £50 million and the United States US$100 million, on condition that the work would be undertaken by British and US consultants and contractors, working with local firms. Later the British funding was topped up by a £100 million loan, subsequently increased to £185 million, arranged by the UK's Midland Bank and Samuel Montagu & Company.

The ODA decided to appoint Taylor Binnie & Partners to continue work on the scheme. Because of the US funding, US consultants were brought in and a joint venture of Black & Veatch from Kansas City and Camp, Dresser & McKee from Boston was selected. The UK and US consultants came together in a consortium known as American–British Consultants (AMBRIC).

The new consortium set up an innovative management structure. There was to be no lead consultant: overall responsibility for the project lay with the Board of Control, comprising one principal from each firm and acting by consensus. The Board of Control met as required, generally quarterly. Day-to-day management rested with a project director and deputy project director, one from the UK and one from the US. There was a fully integrated Anglo-American–Egyptian design team based in Cairo, though some design work was transferred to home offices in the UK and the US. For detailed design and construction, it was decided that the UK would be responsible for the east bank, and the US for the west.

AMBRIC's first step was to review and update the Taylor Binnie master plan whose strategic conclusions it endorsed. The ultimate aims of the project were, first, to remove wastewater from the urban areas of the city and, second, to treat the sewage to a standard to permit its safe reuse in the Nile Delta or for irrigation of desert areas. Maximum use would be made of existing sewers and pumping stations.

The master plan envisaged over 50 km of tunnels (comparable to the length of the Channel Tunnel between Britain and France) on the east bank. The first-phase scheme comprised a new 4–5-m diameter, 12-km long main spine tunnel running more or less south–north, with 3 km of branch tunnel, intercepting the flows from the existing system. Branch tunnels were a maximum of 2.5 m in diameter. The new spine tunnel was to feed sewage by gravity to a new pumping station complex at Ameria, where it would be lifted 24 m into cut-and-cover culverts running 15 km to a new treatment works at Jabal el-Asfar.

But because of the urgency, given the public health risks and the fact that these major new works would take several years to build, the consultants suggested carrying out interim emergency rehabilitation works.

Investigations had already indicated that the 120-km main collector system contained 26,000 m^3 of debris and sediment. Nearly 18 per cent had deposits amounting to more than half the height of the sewer. This was in part because of the original design which adopted shallower gradients and hence lower flow velocities than, in retrospect, would have been ideal.

A pilot cleaning programme indicated that mechanical-bucket cleaning machines were most suitable for removing debris from sewers larger than 450 mm in diameter, and that high-pressure water-jet machines were effective for smaller ones. A major cleaning operation for the main collectors was begun in 1980 in conjunction with the GOSSD. A subsequent closed-circuit TV inspection showed that most of the sewers, except the first collector, were structurally sound. It was also clear, however, that much of the flooding was due to the inadequacies of the local collection system. AMBRIC studied 35 areas and produced pre-design reports for improvements to each. It designed one area in detail as a guide for the remainder, which were then handled by local consultants or the GOSSD.

Detailed inspections of 95 subsidiary pumping stations and the 92 ejector stations, of which 53 were still operational, were carried out. Rehabilitating these stations was considered to be the most effective way of improving the system's efficiency in the short term. (In the long term, construction of the branch tunnels below the existing drains would allow flow from the subsidiary drains to take place under gravity, rendering many of the pumping stations redundant.)

It was decided to replace all mechanical and electrical equipment, including pumps, valves and pipework, in 52 pumping stations. In four areas, where subsidiary pumping stations would still be needed under the final scheme, new stations were built under the rehabilitation programme. Three relief schemes to the overloaded main collectors were also brought forward from the final programme.

Minor improvements were carried out at four major pumping stations which were thought to have limited useful life remaining. More extensive works including provision of new pumps, valves and pipework, were carried out at four others. The activated-sludge plant at Zenein was considered the only one of the existing treatment plants to be worth rehabilitating and keeping in the final scheme. A total of £82 million plus £E33 million was spent on rehabilitation.

Taylor Binnie had concluded that the main sewers of the new system should be tunnelled deep under the city. This was because of Cairo's dense traffic and the potential problems arising from the planned construction of Cairo's Metro. An inverted siphon would have been too likely to fill with grit, while pumping wastewater over the metro would have involved building yet more pumping stations. Tunnelling would also allow most of the existing sewers to discharge into the new system by gravity, reducing the need for local pumping. Branch tunnels would connect to existing sewers just upstream of local pumping stations.

The alternative of a collector along the east bank of the Nile was evaluated and abandoned on the grounds that it would be more expensive and more disruptive to construct. Although access shafts would be needed, a bored-tunnel solution would minimise disruption to Cairo's narrow streets.

Cairo's geology is not ideal for tunnelling. A layer of silt or clay deposited by the Nile is underlain by medium to coarse alluvial sands containing lenses or layers of silt or clay and, below the sands, a thick layer of gravel. The water-table is only 1–2 m below the surface. Underlying this are the Mokattam limestone formations. Tunnelling specialist Mott, Hay & Anderson was employed to help to assess the feasibility of tunnelling, and they jointly concluded that it was technically possible to form tunnels in the overlying alluvial layers.

AMBRIC's US partners were initially doubtful and the alternative of tunnelling at a very deep level through the underlying rock was also considered. The unit cost of a hard rock tunnel was assessed as 30 per cent less than the equivalent in soft ground. This would be offset by the extra cost of sinking shafts to a correspondingly greater depth. Calculations on the trade-off between the two suggested that the rock option would become more expensive if the shafts were between 150 m and 235 m in depth. If the rock head were to be found at 100 m, savings of 3–14 per cent might have been achieved. However from the site investigation it was not clear how deep the rock head was. Boreholes had reached a depth of 80 m without meeting it. In the end the possible savings from rock tunnelling were not judged worth the effort, given the uncertainties of the rock tunnel option.

Tunnels were routed under streets as far as possible to minimise settlement of buildings and the need for land purchase. Several options were considered for building the tunnels. Most of the east bank system was to be constructed in non-cohesive soils below the water-table. The results of the site investigation suggested that stable conditions could be achieved by compressed-air working in most areas, though the larger diameter tunnels were at the limits of size for this technique to be used successfully.

For this reason the design assumed instead that bentonite slurry shields would be used for the northern section 4 m and 5 m tunnels, with compressed air for the southern 4 m spine tunnel and smaller diameters. The slurry technique had the advantage that ground movement would be reduced. AMBRIC was fortunate in having John Bartlett of Mott, Hay & Anderson, recognised as the pioneer of bentonite tunnelling, as adviser.

The main tunnelling work for the east bank project was divided into four contracts for which consortia of British and Egyptian contractors were invited to bid. The Cairo Wastewater Consortium (CWC), comprising Tarmac Construction, Balfour Beatty Construction, Cementation International (now Trafalgar House Construction International), Edmund Nuttall and Egypt's Arab Contractors, won three of them. The fourth went to a joint venture of Lilley Construction and Misr Engineering. Contract starts were delayed by up to nine months partly by the client's cash-flow problems and partly by delays in acquiring the necessary land.

Two of the three pumping stations were built by Kier with the third and largest, at Ameria, going to Christiani & Neilsen with Misr Concrete. Mechanical and electrical work for all three went to General Electric Company (GEC). For the construction of the culverts the stipulation attached to the provision of British finance for the scheme was that only British technical advice should be sought on aspects such as contract management and large-scale concreting and dewatering. Local-led teams therefore carried out these works.

Construction eventually got under way in 1984. But on the first contract the joint venture of Lilley–Misr found itself having to modify its two tunnelling machines before driving had even started. The contractor began by sinking the 11 access shafts along the 4.9-km length of the tunnel which had to be constructed in tight spaces between buildings or in congested streets. AMBRIC's design called for 10-m high reinforced concrete caissons to be sunk under compressed air through the waterlogged silts, clays and sands. The largest was 12.9-m in diameter.

Once the top of the caisson had been sunk to ground level, excavation continued beneath it and the shaft was continued downwards by underpinning it with segmental rings. The deepest shaft went down almost 25 m. The shaft was then bottomed out and lined with reinforced concrete.

Prior to sinking the shafts, Colcrete, now Keller Colcrete, undertook a one-year subcontract worth £3.3 million to stabilise ground around the shafts and where the drives passed close to weak structures. Ground consolidation was achieved by injected chemical or cement/bentonite grout from the surface.

Lilley–Misr's tunnelling machines were made under licence in Britain by Markham to a Japanese design by Okumura. The cutting head had been designed to allow boulders of up to 250 mm to pass through, on the basis of the site investigation which had found boulders of a maximum size of 100 mm. But when the shafts were sunk, boulders of four times that size were discovered. Modified and strengthened heads had to be added to the machines, and they arrived on site six months late, in March 1986.

Once this setback had been overcome, however, tunnelling proceeded smoothly and was completed in 65 weeks, so that the contractor finished two weeks ahead of schedule despite the hold-ups. In the course of the contract one machine set a new world record by driving 168 m in a week, surprising even the machine's designers. The time taken to build a ring, consisting of seven segments, eventually fell to 40 minutes.

Lilley–Misr machines were built to a sophisticated Japanese design, with extensive instrumentation and automation. On the adjacent contract to the north, the Cairo Wastewater Consortium opted for a more basic, but proven, West German design by Bade & Theelan dating from the 1970s. CWC's machines were made under licence by Northern Engineering Industries (NEI) in Leeds in Yorkshire. Some considered the Japanese machine superior because of its better control facilities. But, as Tarmac operations director and chairman of CWC's supervisory board the late David Hoare told *New Civil Engineer*: 'We did not want anything too sophisticated because Egypt has limited servicing facilities for such complex machines.'[6]

The West German machines also suffered initial problems when their rubber tail seals failed, allowing the tunnel invert to be flooded behind the shields. The problem was solved, after a six-month delay, by fitting a new seal similar to that on Lilley's machines, consisting of two wire brushes with fibrous grease between them.

Thereafter CWC's machines performed impressively, regularly exceeding programmed rates, with the contractor making up lost time by starting the brick lining early. The West German-designed machines achieved a maximum drive of 118 m in a week.

For the southernmost section of main tunnel, contract 12, mixed ground conditions with sand containing varying amounts of silt prompted CWC to go for an open shield with backhoe machines working in compressed air at up to 200 kN/m^2. This was satisfactory until the tunnel approached the boundary with Lilley–Misr's contract, where the ground conditions encountered were increasingly unsuitable for the specified compressed air tunnelling. To overcome this problem CWC innovatively leased from Lilley–Misr the Okumura–Markham machine which was driving south of the adjoining contract. This enabled the contract to be completed on time.

CWC used a different but highly successful method of shaft sinking on its contracts, opting to sink the whole depth as a caisson, rather than using the part-caisson, part-underpinning method suggested by the designers. 'Our method was substantially faster with less risk to employee and structure,' David Hoare said at the time. With 27 shafts on its two sections of main tunnel, and 21 on its third contract, for 6.6 km of branch tunnels, CWC was able to turn the process into a virtual assembly line operation. Eventually it was sinking shafts at a speed 25 per cent faster than programmed.

On the smaller 2.5-m branch tunnels the contractor opted for using Canadian-built Lovat earth-pressure-balance machines. The performance of these surpassed all expectations and also achieved driving rates 25 per cent higher than programmed. As a result the branch sewers were finished nine months early.

Difficult ground conditions took their toll on Christiani & Neilsen –Misr Concrete's contract to build the main Ameria pumping station. One of the two 32-m deep concrete caissons stuck midway through sinking and was not freed for 18 months. The main caisson is a double cell structure 45 m in diameter. Wastewater reaching the northern end of the collector tunnel flows first into a 22-m diameter distribution chamber, then through a short tunnel into the annulus between the two cells of the main chamber where it is raised 24 m to the start of the culvert to Jabal el-Asfar. The pumps are housed in the inner cylinder of the main caisson.

But after sinking 15 m the main caisson refused to go any further after the ring of bentonite, acting as a lubricant around its outer face, collapsed. The bentonite ring had to be widened from 150 mm to 600 mm before the caisson would move again.

Meanwhile, the face of the short tunnel between the distribution chamber and the main caisson collapsed when its ground freeze protection failed. Christiani & Neilsen director Patrick Stewart described the contract as 'a nightmare of a job in horrendous ground conditions'. The pump station was completed three years late.[7]

AMBRIC was greatly concerned to protect the new tunnels from the risk of corrosion arising from septic conditions. Accordingly they were all provided with a corrosion-resistant secondary lining of blue bricks, with chemical resin epoxy mortar pointing, above the level which is normally submerged. Unfortunately local brickworks were unable to supply the bricks in the necessary quantities, leading to some delay in completion of the tunnels. On one of CWC's contracts, the bricks were added in a later secondary contract.

Following inauguration of construction by the UK Prime Minister Margaret Thatcher in 1984, and subsequent visits by statesmen and engineers from many countries, the East Bank Phase I scheme was formally commissioned in 1992 by President Mubarak of Egypt and the Duke of Gloucester. It was a matter of pride for those involved that the final out-turn cost was below the total sum tendered. The main priority of ridding Cairo's streets of the health hazard of flooding sewage had been achieved. One significant part of the scheme, however, was not ready namely: the treatment works at Jabal el-Asfar. In the meantime, waste-water has continued to be discharged at Kossous into the Nile Delta drainage system.

Jabal el-Asfar will provide secondary treatment for wastewater flows of 1 million m^3. AMBRIC's original plan had been to build it in modules of 250,000 million m^3/d capacity. After the tunnelling con-tracts had been awarded, however, money was tight. Only £30 million of the promised finance remained. Despite extensive negotiations fol-lowing the submission of tenders, the lowest bid remained £18 million adrift at £48 million. Egypt's creditworthiness had declined, and it proved impossible to find the relatively insignificant missing sum.

While AMBRIC continued to investigate interim solutions, the Egyptian government decided on a new approach and invited bids from European Union member countries for funding and building the entire works. This elicited a proposal from the Italian government with the result that, in late 1991, a consortium of Ansaldo Condotte, Fochi and the NASR General Construction Company started work on a £106 million contract for the works. Completion was expected by the end of 1995.

The treatment works uses a conventional activated-sludge process, intended, like the whole scheme, to be easy to maintain. David Kell says: 'It's designed to be easy to operate to meet Egyptian laws on dis-charge but with the idea that there could be, in the future, reuse of effluent for irrigation purposes, which may or may not require further treatment depending on the particular use'.

Since the commissioning of phase I, further loans from the European Investment Bank have allowed the branch tunnel network to be extended. Contract 15, won by the Cairo Wastewater Consortium, was commenced in 1994. The three and a half-year contract involves the construction of

46 access shafts of up to 6 m diameter, and 10.9 km of 1.8-m and 2.2-m diameter tunnels in 30 different drives under central Cairo. In addition, 37 km of pipeline of up to 1-m diameter are also to be laid. A review of the master plan, which dates in its essentials from 1977, is also being proposed as Cairo's population continues to grow. Studies into the reuse of effluent are also being undertaken.

The challenge of dealing with Cairo's wastewater is by no means over. But the achievement so far of having substantially reduced the threat of water-borne disease is something for which Cairo's population has cause to be thankful.

Provincial water supply

In 1978, while implementation of the Cairo wastewater project had come to a temporary halt, Binnie, Taylor and Dr Abdul Warith successfully bid to develop an improvement strategy for all 24 Egyptian provincial governorates. Three field teams were sent from Cairo to prepare status reports, from which master plans and targets for water supply up to the turn of the century were drawn up. BTE (Binnie Taylor Egypt) then went on to produce detailed plans for Beheira.

The master plan predicted that by the year 2000, the water supply would need to have increased by 65 per cent of its 1978 level to keep pace with a provincial population expected to reach 50 million. The team recommended extensive rehabilitation works nationwide at an estimated cost of US$430 million – US$570 million at contemporary prices. This would cover improvements to source works, uprated chlorination facilities, repairs to reservoirs, and leakage repairs in trunk mains. The ultimate aim was that, by the end of the century, most of the population would have access to a piped supply.

Beheira was typical, with trunk mains carrying treated water for 30–40 km towards the coast where the groundwater was unpalatable because of saline intrusion. As much as 70 per cent of the water was lost through leaking joints, causing standpipes to run dry.

In many villages water was supplied by a ring main feeding standpipes round the perimeter. In large villages, these might be too far away for convenient use by many people. In its proposals for Beheira, BTE recommended installation of 200 km of trunk mains and 400 km of branch mains to remedy this.

Elsewhere, hand pumps were commonly used to extract groundwater, and these could become health hazards because a pool of water would often form around the pump which would then become polluted by grazing animals.

BTE prioritised the regions on criteria such as: expected water shortage by 1990; available alternatives to a public supply; and percentages of rural population without a public supply. Total costs to the year 2000 were estimated at US$1,700 million.

Part of the funding for Beheira was provided by the World Bank, but progress in the Beheira governorate was extremely slow, mainly because of a lack of local finance. However, four new water treatment plants with a total capacity of 1,900 l/s were due to be completed in 1995. The master plan was financed in the other provinces mainly through bilateral aid.

Roads

During 1975 the British consultant Ward, Ashcroft & Parkman was involved, with Parsons, Brinckerhoff, Quade & Douglas of the US, local consultant Sabbour Associates and the Egyptian General Organisation for Physical Planning, in the Cairo Entrances Study. It had the purpose of making recommendations for roadway entrances to Greater Cairo which would take account of planned new developments within the region, such as those in the Suez Canal Zone, and in Egypt as a whole.

The study was an extension of two earlier plans: the Preliminary Master Plan for the Greater Cairo Region (1970), and the Cairo Transportation Study (1973). These proposed a highway system based on an outer ring road, an intermediate ring road, and a system of radial and other highways. The outer ring road was to act as a boundary to future urban development, and three satellite settlements were proposed outside the ring road. The study was completed in eight months.

Construction of Cairo's new roads has demonstrated local expertise. The system of elevated highways has vastly improved traffic flow in Cairo. They were constructed on a design-and-build basis by Arab Contractors with support from local consultants.

There have been several new crossings of the Nile, of which the first was the Sixth of October bridge, built in the mid-1970s. The eight-lane bridge was 2.7 km long with 3.5 km of slip-roads. Documentation and steel for the prestressing system, supplied by the West German firm Dywidag, was the only non-Egyptian content.

Egypt today

Without substantial oil reserves, Egypt has not enjoyed the prosperity which the other oil-producing states experienced. Instead it has suffered from continuing economic crises and high levels of debt which in 1990 reached US$50,000 million, more than the country's GNP. Following lengthy negotiations with the IMF and international creditor governments, the write-off of half Egypt's debt in stages over three years was agreed in 1991. The second phase of this programme was agreed in September 1993. Egypt's government has begun the process of privatisation of many industries brought under state control in the Nasser era. Domestic debate centres around this issue. Some argue that Egypt's state bureaucracy stifles economic development and that, in consequence, privatisation should go faster, while others believe that state control helps to protect the Egyptian population against the vagaries of private enterprise.

Meanwhile re-establishment of diplomatic links with other Arab states, broken after the Camp David accords with Israel in 1978, have resulted in the resumption of aid from Kuwait and Saudi Arabia. Nevertheless, it is likely to be some years yet before Egypt finds itself on a more secure economic footing. In these circumstances, construction projects on the scale of the Greater Cairo Wastewater Project are likely to continue to depend on outside aid.

One of the most charismatic and influential figures in Egyptian construction in modern times was Osman Ahmed Osman. He founded a small contracting firm in the Suez Canal zone after World War II, and had to start at the bottom twice after amassing and losing fortunes. His reputation was secured when his firm Arab Contractors took over and completed construction of the Aswan High Dam during the 1960s, after other local firms had fallen foul of the Nile geology and the demands of the scheme's Soviet backers. He became Minister for Housing and Reconstruction at the time of the Suez isthmus reconstruction, with the brief to attract international industry to the area. His responsibilities also included the renewal of public utilities in Cairo – water-supply, drainage and roads. He went on to become deputy prime minister in 1981.

As a minister he remained first of all a businessman. 'I hate bureaucracy. I hate to be an employee. I am a free man,' he told *New Civil Engineer* in 1974. 'Liberty is the most important thing for a businessman, and not only for a businessman, but for any work to be done perfectly you should have freedom in your movements, in your thoughts, in everything. But if you give yourself or your business to bureaucracy, you will not finish anything. So I am here in my ministry to kill bureaucracy.'[8]

chapter fourteen

JORDAN

The Kingdom of Jordan became an independent state in 1946.
Although it has more limited natural resources than its oil-rich neighbours, its assets have included a well-educated and skilled population,
and for many years remittances sent home by thousands of Jordanians
employed in Gulf states have bolstered the country's balance of
payments.

Jordan's short history has been inextricably bound up with the
Arab–Israeli conflict. This has greatly influenced the development of its
economy. The loss of the West Bank of the River Jordan to Israel in
1967, for example, deprived Jordan of an important area of agricultural
production. The influx of large numbers of Palestinian refugees has also
made job creation urgent.

The people of the land which is now Jordan became predominantly
Muslim after the defeat of Byzantine forces in AD 636. Between the
eleventh and thirteenth centuries, the Crusades sought, unsuccessfully,
to wrest Jerusalem and the Holy Lands from Islamic rule. The region
became part of the Ottoman empire in the sixteenth century; Ottoman
rule lasted for four centuries until World War I. After this war the area
then known as Transjordan (east of the River Jordan), became part of
Britain's Palestine Mandate. In 1923 Britain granted Transjordan recognition as an independent state ruled by Amir Abdullah; however it
remained in reality, subject to substantial British control under the mandate. A treaty five years later extended the state's independence but stated that Transjordan would be guided by the advice of the British
Resident in financial and foreign policy.

Full independence was granted in 1946 when Abdullah was
crowned king of the Hashemite Kingdom of Jordan. However the treaty
granting independence allowed Britain to continue to have military
installations there.

The first Arab–Israeli war, which broke out immediately after the
British mandate ended in 1948, left Jordan in control of the West Bank,
the area of east Palestine which had been designated to be an Arab state
under the United Nations partition proposals. Jordan also found itself

having to cope with 400,000 refugees. The West Bank was formally incorporated into Jordan two years later; following the 1967 war it was occupied by Israel, and has remained so since then. Jordan has since abandoned all claims to the territory in favour of the Palestinians.

Following the assassination of King Abdullah in 1951, the short reign of his son Talal was notable mainly for constitutional reforms making the cabinet accountable to parliament. Talal abdicated after a few months and was succeeded by his son, Hussein, then still in his teens and who had been educated in Britain. Hussein's reign began uncertainly, with phases of civil disorder. One of his early decisions was to allow free multi-party elections, but clashes between the resulting radical parliament and the monarchy, and a suspected coup attempt in the mid-1950s, caused Hussein to ban political parties and set up a more authoritarian regime. It was 1993 before multi-party elections were again held.

One initiative which the radical parliament of 1956 did succeed in carrying through was a negotiated end to the treaty with Britain. Hussein then established links with the US, which replaced Britain as the chief source of foreign aid.

The Arab–Israeli war of 1967 resulted in a crushing defeat for Jordan, Egypt and Syria. Jordan lost control of the West Bank including Arab Jerusalem; this was a blow not only to its self-esteem but to its economy, leading to a sharp decline in agricultural production. Irrigation of the Jordan Valley with waters from the Yarmouk, a major tributary of the River Jordan, has failed to compensate for this.

Jordan benefited from the oil boom of the 1970s, to a large extent through remittances sent home by about 350,000 Jordanian nationals working in other Middle Eastern states. But the recession of the 1980s and the Gulf War had their effect, causing a significant decline in income.

The prosperity of the 1970s nevertheless left Jordan with an efficient and modern infrastructure, and with some importance as a centre for commerce. However, the country did not build up a strong industrial base. There are a few medium-sized manufacturing concerns, alongside large industries in which the state has a majority shareholding. These include those of phosphate and potash production and related down-stream industries such as fertiliser manufacture. Extraction of potash from the Dead Sea has been one of Jordan's major projects since the late 1970s. Small amounts of oil have been discovered.

The Port of Aqaba

In the early years of independence during the 1950s, Jordan had few sources of income. Development of its phosphate reserves for export was therefore seen as crucial for the country's economic survival.

Phosphate deposits had been discovered in Rusaifa, near Amman, by Amin Kawar, a Palestinian, in the 1930s, and he had exported them from the Mediterranean port of Haifa. With the expiry of the British Mandate, access to Haifa was denied. Kawar's son Tafiq sought to develop new phosphate mines at El-Hassa, between Amman and Ma'an in south Jordan. King Hussein supported the idea and in 1953 the Jordanian government doubled the capital of the Jordan Phosphate Mines Company and became a major shareholder in it. At the same time King Hussein decided to develop Aqaba for phosphate exports.

Jordan is almost land-locked – its only coastline is an 8 km stretch along the Gulf of Aqaba. Some 3,000 years ago the port of Ezion-Geber was sited there. It was a flourishing centre where caravans from Damascus and the interior met and traded with ships sailing up the Red Sea from Africa and the East. As trade routes changed over the centuries, however, its importance declined.

In late 1952, when Rendel Palmer & Tritton was commissioned by the Jordan government to prepare a master plan for development of the port of Aqaba, there were few facilities there. A small lighter basin had been built by the Allied forces in World War II which was still used for bringing in military stores. This was at the north of the town; to the south was a short lighter wharf used for civilian imports.

Most of Jordan's imports, around 100,000 t/y, then entered the country via Beirut in Lebanon. Rendel Palmer & Tritton estimated that 60,000 t/y could be expected to transfer to Aqaba. The main traffic through Aqaba would be phosphate exports, which Rendel Palmer & Tritton estimated could reach 100,000 t/y within 10 years. On this basis it recommended that a deepwater phosphate berth should be built, but concluded that a deepwater general cargo berth would not be justified and that a lighter wharf should be provided.

The government decided to go ahead with a deepwater berth for general cargo in any case; this proved justified when the levels of imports and exports each individually reached 100,000 t/y by 1960.

The site chosen, south of the town, had deep water close inshore and an area of land adjacent for development. Provision of the customary sheltered area for shipping was impossible because the shore shelved too steeply to build a breakwater, and extensive rock excavation for an inland basin was not economic. Open-sea berths therefore had to be provided. Because the prevailing wind for 10 months of the year is from the north, the berths were aligned almost due north–south. It was estimated that storms from the south, which occur in winter, would render the berths unworkable for only 10 days a year.

The nearest railhead was at Naqb Ishtar, 42 km south-west of Ma'an and still 80 km from the port. The Hejaz railway, which ran from Damascus to Medina, had been derelict from Ma'an since World War I, when it had been largely destroyed by Arab troops fighting under T. E. Lawrence and Sharif Hussein of Mecca, but had been extended to Naqb Ishtar in 1942 by the Allies. A road had been built to cover the remaining distance from Naqb Ishtar to Aqaba. Since it was considered unlikely that it would be economically feasible for some considerable time to extend the railway, the new port was designed with wide roads, to be served by lorries. However provision was made for rail to serve the phosphate stores if this should become necessary.

In a paper to a 1960 ICE conference on 'Civil Engineering Problems Overseas', Rendel Palmer & Tritton's partner John Palmer said: 'Few heavy civil engineering structures had ever been built in Jordan, and the small cadre of skilled craftsmen had little inducement to move down to so remote a spot [as Aqaba]. The overriding principle in the design of the works was therefore to make them as simple as possible.'[1]

The cargo berth has an 18-m wide deck carried on steel box piles with its front face in 10 m of water, and a sheet-piled retaining wall at 5-m depth. It was designed to be used by any vessel from the size of a lighter to a ship of up to 20,000 dwt. The phosphate berth was designed like a conventional oil berth with two berthing dolphins and a loading tower between. There is a 110-m x 60-m transit shed which includes a control tower and offices in one corner.

Two phosphate stores of 20,000-t capacity were provided, with mechanical loading plant of 500 t/h capacity. A 1,300-kVA electricity generating station was built to provide power for the port as well as 100 kVA for street lighting in Aqaba.

The contract was awarded to Ed Zueblin of West Germany in October 1956, but the start of work was delayed for almost a year by the Suez crisis. The first phosphate ship was loaded two years later in August 1959. The cost of the port was just £2 million.

A new highway was also provided linking Amman with the new port, a distance of 200 km. The road, which cost £3.6 million, included seven bridges. It was largely completed in 1960, though final surfacing was not carried out until three years later.

Traffic through the port increased in line with the more optimistic forecasts and within a few years an extension was justified. Provision for extra berths had been included in the original plans and was put in hand in 1963. A second 180-m long cargo berth was added, as was a 7,000-m^2 transit shed, and 4,200 m^2 of covered storage.

In the early 1970s it was deemed economic after all to construct a railway from the mine at El-Hassa to Aqaba. The 270-km line was opened in 1975. Between El-Hassa and Ma'an the trackbed of the Hejaz railway was used, with a new extension from Ma'an to Aqaba. It was extended 25 km to Manzil in 1980 and by a further 27 km to new mines at Wadi al-Abyad two years later. Rendel Palmer & Tritton was commissioned to design and supervise construction of the phosphate handling system. Work included rail sidings, discharge hoppers, conveyor systems, storage sheds, and a reclaiming system and shiploader with a capacity of 2,100 t/h. At Wadi al-Abyad the consultant prepared an outline design for a wagon-loading plant capable of loading two wagons simultaneously.

The reopening of the Suez Canal in the mid-1970s led to increased growth in traffic through Aqaba, and the Jordanian authorities sought to expand the port further. Because of a general desire to introduce more competition, the commission for a new master plan was put out to competitive tender. The job was won by the UK firm Parsons, Brown & Newton, a port specialist whose traditional area of work was the oil industry. It had been seeking new markets because, in the second half of the 1970s, a decline in its oil workload was becoming apparent, the firm has since merged with Mott MacDonald.

Parsons recommended construction of two large temporary berths to relieve congestion in the short term, while seven additional permanent berths were built. The £20 million permanent works included four new general cargo berths 180 m long, in 12 m of water and capable of handling vessels of up to 40,000 dwt; two ro-ro berths for ships of up to 8,000 dwt; a 300 m wharf for coastal vessels of up to 300 m; and a 150-m extension to the existing lighter wharf.

'Rendels had put the first wharves in the best place, where the land was near deep water,' says Mott MacDonald director Mike Blackburn, who worked at Aqaba as resident engineer. 'We had to reclaim a big patch of land over coral reefs.' The new jetties were built of tubular steel piles at 4.5-m centres. The front row of piles was spanned by a precast concrete trough with an in-situ berthing beam. Precast U-troughs also spanned along rows of piles at right angles to the berthing beam, and the deck was created by prestressed concrete planks supported on these troughs, with a shallow in-situ topping. Behind the piles the reclaimed land was formed, with an armoured slope.

The usual associated facilities, including transit sheds, open and covered storage, a cold store and a container park were also constructed. There was provision for container cranes to be added later. The UK construction company Tarmac was the contractor for the work, which was completed in 1980.

The temporary berths were enormous pontoons towed from Japan and positioned 5 km south of the existing port. They were supplied with cranes and access bridges serving cargo handling facilities on the shore, and were capable of handling cargo ships of up to 15,000 dwt and ro-ro vessels of up to 7,000 dwt. The temporary berths were fabricated and installed within 12 months, in 1978. The cost of the works was £33 million at 1980 prices.

Midway through the works the Iran–Iraq War broke out and Aqaba became a main supply point for Iraq. The berths were used to land supplies of munitions, and the lorries started to wreck the roads. These munitions' lorries were often grossly overloaded – weighing up to 100 t and left the road to Amman heavily rutted.

With the inception of the Dead Sea Potash project in the 1970s, a further terminal and jetty was needed for export of potash. Parsons was again responsible for designing a piled jetty with a partly precast and partly in-situ concrete deck. A transshipment store, added part way through the project, and its handling facilities, were designed in association with Rendel Palmer & Tritton. One million m^3 of soil was excavated and re-deposited to create a level site platform from the steeply undulating desert.

The transshipment store is a 350-m long twin portal-frame structure, each half spanning 56 m. It is capable of holding 150,000 t of potash, which is delivered by truck and stored until a shipload has accumulated. Potash can be loaded into ships at the rate of 2,000 t/h. The main contractor for the terminal was PHB Weserhutte; Wimpey was subcontractor for construction of the shed, with structural steel supplied by another British firm Cleveland Bridge. Ed Zueblin built the jetty. Ancillary works included provision of weighbridges and conveyors, administration buildings, workshops and services. The terminal was completed in 1982.

More recently Rendel Palmer & Tritton has been reappointed at

Aqaba, and designed a new terminal for the export of oil. The terminal consists of a 150-m long piled approach arm and jetty head with berthing and mooring dolphins. The approach arm consists of 15-m long structural steelwork spans carrying pipeways and a concrete roadway, while the dolphins are made of concrete slabs supported on tubular steel piles. A 400,000-dwt oil storage tanker or 'floating tank farm' is permanently moored to the terminal, and ocean-going tankers berth alongside to be filled. Onshore is a 20,000-m^2 paved area connected to the highway where road tankers delivering oil are connected to a pipeline through which their load is pumped to the storage tanker.

The Dead Sea Potash project

The Dead Sea lies 400 m below sea level, the lowest body of water on earth. The River Jordan flows into it but nothing flows out – water is lost only through evaporation. The sea has amassed high concentrations of mineral salts – in particular, sodium chloride or common salt, magnesium chloride, and potassium chloride (from which potash can be extracted). Potash is used to make fertilisers, and the Dead Sea is estimated to contain as much as 1,000 million t of it.

In the mid-1960s the first studies were carried out into the feasibility of recovering this potash. The client, The Arab Potash Company, appointed Jacobs International of California as project management consultant and process designer; Jacobs brought in Sir Alexander Gibb & Partners to advise on civil engineering aspects and Wimpey Laboratories won the site investigation work.

It was concluded that, with its low rainfall, low humidity and high temperatures, the area around the Dead Sea provided ideal conditions for recovering potash by evaporation. During the 1967 Six-Day War the area became a front-line zone and a plan to produce 500,000 t/y was put on hold. Israel had started work on its own potash scheme at about the same time on the West side of the Truce Line and this went ahead.

It was not until 1975 that the Jordanian plan was revived. The way in which potash is produced is by pumping Dead Sea brine into large, but relatively shallow, evaporation pans enclosed by earth dikes, where the concentration of salts increases until they precipitate out. The specific gravity of Dead Sea water is 1.22; sodium chloride crystallises and precipitates out at a specific gravity of 1.26. The water is then transferred to further pans where evaporation continues until, at a specific gravity of

1.31, carnallite, a double chloride salt of potassium and magnesium, is deposited. This is dredged from the bottom of the pan and pumped to a process plant where the potassium chloride is separated out.

Production capacity is determined by the area available to build evaporation pans. The Dead Sea consists of two basins, the north (up to 350 m deep with steep shores) and the south (shallow with a flat shore-line). The level of the Dead Sea is dropping by around 0.8 m a year because of climatic effects and abstraction of water from the rivers flowing into the Dead Sea basin, principally the River Jordan. In between the first and second feasibility studies the south basin had shrunk until there was very little water left in it. This had left substantial salt flats around it, making it possible to build more extensive evaporation pans than originally planned, with water extracted from the deep northern basin. The revised scheme envisaged a £224 million project with an initial production capacity of 1.2 million t/y, representing around 3 per cent of world potash output.

Wimpey Laboratories again won the site investigation contract for the 240 km^2 area, and ran into various problems, some foreseen and some not. It had been known that the area to be investigated would be covered with soft mud washed in from the surrounding rivers and wadis. Specialist plant was accordingly used. The drilling rig was manœuvred around on a Mackley Ace hoverpontoon, towed by a Gemco amphibious transporter. A Nimbus hovercraft was used for reconnaissance. It had been intended to use a launch to do the towing, but it proved impossible to bring it from the north basin because of the discovery of unsuspected salt reefs. Towing with the Gemco was slow, and jagged salt formations left behind by evaporation caused considerable damage to the skirts of the hovercraft.

Wimpey would have liked to use a helicopter, but the Jordanian military authorities refused to allow this. But the most serious unexpected problem was the discovery that the area had been mined. The existence of mines had not been suspected until the first one was found. Wimpey talked to the Jordanian authorities. 'There was some mention of the wadis which flood into the Dead Sea being mined at one time,' Wimpey's project manager Malcolm Gamble told *New Civil Engineer* at the time. 'Nothing was done; everyone kept their fingers crossed that the mine we found was only a one-off.'[2]

Then more were found, and a Jordanian army unit was moved in to clear the area. They found about 1,000 mines. Afterwards all went well until near the end of the investigation when yet more mines were discovered near the outfall from the Israeli potash scheme on the opposite shore. At this point the army finally agreed to provide a helicopter to allow Wimpey to finish the job.

There were two main engineering problems involved in building the evaporation pans, says Gibb director Tim Woods Ballard. First, he says: 'It was a substantial engineering challenge building dikes to hold brine in some of the softest mud ever recorded.' The second problem was preventing seepage through dikes and through the layers of salt which lie over the site alternating with the layers of mud. This was most serious for the western perimeter dike, where any seepage would effectively be lost to the system. For the dikes between pans, seepage would be undesirable but not critical, provided it did not affect stability.

The site investigation confirmed the suitability of materials existing on site for construction of the dikes, including the unique Lisan marl for the cores.

There are 65 km of dikes, of which 45 km are supported on the soft silty clay. Because of the nature of the clay, and the inherent limitations of site-investigation techniques, three full-scale trial dikes were built with differing construction methods and cross-sections and loaded to failure.

Apart from the possibility of failure, large settlements as well as initial sinking of fill material into the ground were expected. For a 3.5 m high dike, the average settlement was predicted to be one metre over the construction period, with a further 0.5 m over the next 10 years. Construction of the dikes in several stages, with a gap of several months in between, allowed the foundation material to consolidate and gain strength, limiting settlement and reducing the loss of material into the foundations as well as the risk of failure. Completion of roads and pipe runs along the top of the dikes was left for a year after completion to allow settlement to be made good.

The site investigation had revealed large variations in foundation strength over short distances. Designing for the lowest strength throughout would have been uneconomic, so the 90 percentile strength was used at any given depth. In theory this meant that up to 10 per cent of the total length of dikes could fail. But again, staged construction allowed this to be avoided. Factors of safety of 1.2 were used for the construction condition and 1.5 for the final state, but the construction stage was found to be more critical.

Calculations showed that, with a marl core, seepage through the dikes would be small. Potential seepage through the foundations was significant, however, particularly where crystalline salt deposits existed. A 15-km length of western perimeter dike was therefore provided with a vertical cut-off wall under the core. This was constructed using a technique devised by Wimpey International, which won the £65 million contract for building the Solar Evaporation System. A 225 mm wide slot 6 m deep was formed using a specially devised trench cutting machine fitted with a coal-cutting saw, which resembled an enormous chain-saw. At this depth the cut-off trench reached an essentially continuous clay layer below. A semi-rigid high-density polyethylene impermeable membrane was placed in the slot and supported with a self-setting clay/cement/brine slurry.

The dikes are up to 60 m wide overall including berms. Their height is determined by the minimum depth of brine needed, the amount of salt and carnallite deposited and estimated post-construction settlement. Construction involved excavating 1 million m^3 of mud and depositing over 8 million m^3 of fill. The berms are made of gravel, with a sand and marl core. Slopes are protected from wave erosion by a geotextile membrane and rip-rap. It was expected that after 10 years or so some of the dikes would have to be raised to accommodate rising deposits of sodium chloride on the bottom of the pans.

The sequence of construction for the dikes adopted by Wimpey began with the end tipping of two parallel bunds of bulk fill gravel. This made it easier to remove mud and salt slush beneath core level between the bunds. Where no cut-off wall was required, the next step was to place the core, surrounded by fine filter sand, as soon as excavations reached a suitable foundation. Where a cut-off wall was needed fill material was placed between the bunds to make a floor for the necessary plant. Once the cut-off had been formed, the dikes could be raised in stages to their finished height and be compacted with smooth-wheeled vibratory rollers.

The main brine intake is built on a piled jetty in the north basin, 11 km from the main salt pan. In summer, when evaporation is at its highest, pumping can reach 4,300 m^3/h. A canal delivers the brine into the salt pan where unwanted sodium chloride first precipitates out. Further pumping stations transfer the brine into 'pre-carnallite' pans where further concentration takes place, and finally into the three carnallite pans themselves. Carnallite is continually dredged using floating harvesters once it has precipitated. 'Control of the specific gravity of the brines through the pan system is a great art.' says Tim Woods Ballard.

So that the system was ready to produce potash immediately the 39-month construction contract ended, Wimpey had to meet a number of staged completion dates, each of which had associated bonuses and penalties. The contract had been signed in February 1979. By 1 May 1980 pumping of brine into carnallite pan C1 had to begin, to make a start on the formation of a salt lining to the bottom of the pan. An initial layer of sodium chloride was required on the soft mud surface of the base of the carnallite pan to enable the carnallite to be deposited on and then harvested from a firm base. This meant that a temporary pumping system had to be installed within 15 months of the start of the contract and involved providing 18 km of twin 1.2-m diameter GRP pipeline and a floating intake station.

By February 1981 delivery of brine to the main salt pan had to begin, requiring completion of the permanent main brine intake jetty, and completion of dikes around the salt pan to an appropriate level. By the following month brine had to be transferred to the pre-carnallite pans, and a start made on laying the salt floor of carnallite pan C2. Again this meant that the dikes had to have reached minimum levels and that the brine-transfer pump stations serving pans PC1, PC2, C1 and C2 had to be ready. Final completion date was 31 March 1981. The contractor succeeded in meeting all these deadlines.

The project is in a remote area and a township was constructed, eventually to accommodate a population of 3,000. It was built by a Korean contractor before the main contract was started so that it could be used by the construction personnel. Wimpey won further contracts to build a hospital, mosque and guest-house and had responsibility for operating and maintaining the township during the contract.

Since completion of the Solar Evaporation System in 1982 the continued fall in the level of the Dead Sea has had two effects: the intake is now due to be repositioned, and a new beach has been exposed off the Lisan peninsula.

Since the project came on stream many new projects have been undertaken by the Arab Potash Company. Output has been raised to 1.8 million t/y by optimising the solar evaporation system, and by adding additional processes in the refinery to extract more potash from the precipitated carnallite. Due to the deposition of sodium chloride in the salt pan the enclosing dikes have had to be raised twice, each time by about 1.5 m. A feasibility study has been completed to determine whether it is economically worth while to expand production further by building a new salt pan on the Lisan beach. A further carnallite pan was constructed near the refinery in 1995. These additions could increase output to 2.2 million t/y, probably the upper limit given the physical constraints of the site, Tim Woods Ballard believes. The Lisan beach slopes rather more steeply than the original site so that the size of pan which can be enclosed by a dike of a given height is much smaller.

Water supply and sewerage

The Great Yarmouk project was primarily intended to provide irrigation water for agricultural land in the lower Jordan Valley. But following a four-year drought in the late 1970s, it was decided to divert some of the water to the capital, Amman.

Stanley Consultants and Boyle Engineering of the US designed the £80 million scheme. The first stage involved pumping 60 million m³ of water from the East Ghor canal to a service reservoir near Amman, 1,300 m higher. Here it passes through a treatment plant before being pumped again to a reservoir holding two days' supply. Stage II doubled this supply. Finance was provided by the US Agency for International Development, and by Saudi Arabia which financed the pipelines and pumping stations.

Almost a decade later, water supplies to the capital were further supplemented from a groundwater source in the Wadi Wala, to the south-west. John Taylor & Sons, with the local firm Ruqn al-Handasa, was appointed to review existing designs and supervise construction.

The works comprised a wellfield in the wadi producing 68 million l/d, 100 km of buried steel pipeline, three reinforced concrete service reservoirs of up to 12,000 m³ capacity, and a telemetry control system. Again Saudi Arabia contributed to the £19.8 million project's cost.

At around the same time, in the early 1980s, the sewage treatment works serving Amman was seriously overloaded. A quick but economic solution was sought. Because space on the existing site was limited, it was decided to take the flow to a remote desert site 40 km away. Levels were such that flow could be by gravity through an inverted siphon.

Binnie & Partners was commissioned by the Turkish contractor Gama Endustri Tesisleri to carry out detailed design. Design work began in June 1983, and the works was constructed within two years. The £30 million plant serves a population of 750,000. The 40-km inverted siphon is 1,200 mm in diameter and follows the route of a wadi as far as possible, avoiding private land. The treatment works is one of the largest of its kind, and is designed to treat the wastewater to a standard high enough for unrestricted use in irrigation.

Jordan today

In recent times Jordan has suffered as a result of its support for Iraq in the Gulf War. Because of its economic links with Iraq, Jordan was also affected by the imposition of UN sanctions against Iraq. In addition much international aid was also withdrawn. There was some compensation for this in that Jordan's economy was unexpectedly boosted by investment brought by Jordanians expelled from Kuwait. A rapprochement with the US came relatively quickly after the war, though the same cannot be said for Jordan's relations with Saudi Arabia.

Two major problems face Jordan in the immediate future. One is the implications of the 1993 Middle East Peace Agreement: Jordan has long been in favour of a negotiated settlement but had hoped for an accord which would boost its economy through integration with that of any autonomous Palestinian state which emerged. It is not clear that this hope will be realised. Second is the problem of water: under a 1953 agreement, Jordan is entitled to 720 million m^3 a year from the Jordan and Yarmouk rivers while Israel is entitled to 567 million m^3. In fact, Jordan receives only 100 million m^3 a year while Israel takes over 700 million m^3. At the moment Jordan uses only 45 m^3 of water annually per head. But with its population expanding rapidly, conservation of water and the development of new supplies are likely to become more important.

c h a p t e r f i f t e e n

TURKEY

Turkey is outside the definition of the Middle East which this book has adopted. Since World War II it has seen its political future as part of Europe, with which it has sought closer links. It is not an Arab country – less than one per cent of the population named Arabic as their mother tongue in the last census.

Nevertheless, Turkey has had strong historical connections with the Middle East for over two millenia. Such links were particularly strong when much of what is now known as the Middle East was part of the Ottoman empire. Today it continues to have important trading links with its Middle Eastern neighbours; it is also the overland gateway to the Middle East from Europe, and many imports to the Arab world reach their destination by crossing Turkey.

Turkey's historic role on the Europe–Middle East trade route looks set to increase in importance. The two bridges across the Bosporus, and Turkey's huge programme of motorway construction now in progress, will exploit the country's geographical position. For this reason it is considered appropriate to include here a chapter on engineering developments in Turkey.

The history of Turkey as a country can be traced back to 1071, when the Byzantine emperor, Romanus Diogenes, was defeated by the Seljuk sultan, Arp Arslan. Muslim Turks then began to settle in Asia Minor, or Anatolia. The name 'Turkey' was first used about two centuries later.

In the wake of the Mongol invasion which overthrew the Seljuk Sultanate in 1243, a number of Turkish warrior princes emerged who gained control of Asia Minor. One of these, Osman, came to power in 1280. He and his dynasty set up an empire which grew to cover most of the Muslim world – the Ottoman empire. The empire lasted for four centuries, went into decline in the nineteenth century and was finally defeated in World War I.

After the war the Allies proposed to dismember Turkey under the Treaty of Sèvres but were prevented from doing so by Mustafa Kemal, later known as Atatürk, who led the nationalist resistance movement there. After two years of war, the nationalists defeated a Greek invasion force, and forced French occupation troops to withdraw from the south. Kemal's de facto government, set up in 1920, was recognised in the Treaty of Lausanne, and the Republic of Turkey was founded.

Kemal remained president of the republic until his death in 1938. Under his rule Turkey's institutions were secularised and westernised, completing a process which had begun in the last stages of the Ottoman empire. Among the changes which Kemal saw through were the adoption of many laws from Western Europe, the use of the Latin alphabet instead of Arabic script, and the imposition of European forms of dress.

The economic policies of Kemal and his Republican People's Party (RPP) encouraged the emergence of a commercial class, and expanded the public sector in response to an economic crisis in 1929. Kemal's policies also introduced the longest period of peace in Turkey's history. Meanwhile, the republic's first factories were built and the railway system extended.

Turkey remained neutral in World War II. After the war, faced with a demand by the Soviet Union for the return of certain land in eastern Anatolia, the republic sought support from the West. It signed a military assistance agreement with the US in 1947, joined the Organisation of European Economic Cooperation (later known as the Organisation for Economic Cooperation and Development) in 1948, and the North Atlantic Treaty Organisation (NATO) in 1952. An association agreement with the European Economic Community was signed in 1963; Turkey applied for full membership of the Community in 1987, but its application was not accepted at that time.

Atatürk's successor, Ismet Inonu, oversaw the transition from sole party rule to a multi-party state in the late 1940s. The Democratic Party (DP), formed out of discontent with the RPP because the latter appeared to rule on behalf of a narrow élite, won an absolute majority in the national assembly in the first free elections in 1950.

Turkish politics since then have been characterised by free elections interspersed with military take-overs triggered either by economic crises or a breakdown in law and order. Periods of military rule have generally been relatively short, but the return of power to civilians has typically been accompanied by amendments to the constitution and the dissolution of one or more political parties.

The 10 years of DP rule, from 1950 to 1960, were years of unprecedented growth. The road network was built up, largely under US guidance, and a dam-building programme begun; the area under cultivation grew by 50 per cent and production of cereals doubled; industrial growth averaged over 8 per cent a year.

This period ended with increased inflation and a devaluation of the currency; but, after a military interregnum, growth resumed with a series of five-year plans beginning in 1963. But this time pursuit of growth ran into problems after the oil price increase of 1973, when pursuit of the same policies caused runaway inflation, a record balance of trade deficit, and a decline in GNP in 1979 and 1980.

A healthy rate of growth was restored in the 1980s – following another three years of military rule. Turkey was self-sufficient in food but was still far from self-sufficient in industrial goods. Since the mid-1980s Turkey has carried out a programme of expansion of hydroelectric and thermal power production to overcome a chronic power shortage. Plans to develop nuclear power have not come to fruition. There has also been a huge motorway building programme – some 1,550 km of motorway was finished or under construction by 1991, compared with a network of only 84 km in 1984, and it was planned to double this by the end of the century. Meanwhile, a project known as GAP was launched to stimulate new industry and open up southern Anatolia.

The Bosporus bridges

Istanbul, Turkey's former capital and still its largest city, lies partly in Europe and partly in Asia, divided by the Bosporus strait. The western, European, side is the centre of business and commerce while the eastern side is largely residential. For hundreds of years construction of a fixed crossing had been contemplated but it was not until the second half of this century that this became genuinely practicable.

The first serious study was carried out by the US firm De Leuw Cather in 1956. It concluded that a crossing was feasible and economically viable, and identified an alignment for a suspension bridge. New York's Steinman, Boynton, Gronquist & London was subsequently commissioned to produce a detailed design for the bridge. This had a central span of 942 m, with headroom of 50 m, and carried four lanes of traffic. Both towers were positioned in the water. Although it was ready to go out to tender in 1960, political unrest caused the project to be shelved.

The scheme was revived in 1967 when it became clear that ferry services across the Bosporus could no longer cope with the volume of traffic. In 1960, the US had been the only nation to have experience of building long-span suspension bridges. But since then, with the construction of the Forth Road Bridge in Scotland, the Severn Bridge in between England and Wales and the Tagus bridge in Portugal, European expertise had been acquired. Turkey had also established links with the EEC so that European Investment Bank (EIB) funding was potentially available. Proposals were therefore invited from consultants in Europe as well as the US with the result that Freeman Fox & Partners of the UK was appointed consultant on the project in 1968.

Congestion on the ferries meant that the bridge was now urgently needed and the original programme envisaged that it would be designed in nine months and built within four years. This was a remarkably ambitious programme for what was to be the world's fourth-longest suspension bridge. Although there was some slippage, the bridge was indeed completed in record time.

Building on the experience gained during construction of the Severn Bridge, Freeman Fox & Partners' preferred design employed an aerodynamically profiled box deck just 3 m deep. Tenderers were offered the option of bidding on an alternative lattice girder deck which used more steel but which, it was thought, might lead to lower prices overall because of some contractors' greater familiarity with this type of design. In the event, there were no offers for this scheme.

Freeman Fox & Partners' design had a main span of 1,074 m which was longer than that in the US version, and it had both towers sited on dry land. This was considered to pose less risk during construction and gave a reduction in both costs and timing. It also meant that the contractor had to be mobilised and ready to start building the super-structure – which included fabricating a large amount of special equip-ment – a year sooner than would normally have been expected. The headroom for shipping was increased to 64 m, while the number of traffic lanes was raised to six. The extra lanes increased the cost by only 16 per cent and were expected to cater for traffic growth well into the future. The alignment, forming part of Istanbul's peripheral highway, was already fixed by the position of existing roads.

Unusually, though not uniquely, the side spans are not suspended from the main cables. The banks of the Bosporus slope steeply from the shore, and foundation conditions were suitable for the side spans to be built independently as a conventional viaduct. This was estimated to save US$1.2 million in construction costs; it simplified construction of the anchorages and eliminated problems connected with erecting sus-pended box sections over land. Perhaps, more importantly, it allowed further savings in time because work on the approach spans could progress independently of that on the suspended spans.

Freeman Fox & Partners' design also proposed suspension cables of pre-formed parallel wire strands in preference to the more conventional aerial spinning technique, which only a limited number of firms had the expertise to handle. In fact, the winning tenderer offered a saving of £1.67 million to erect the cables by aerial spinning, considering that the alternative was unproven at the time, and this was accepted. The bridge was also designed to resist earthquake loads.

Tendering had been delayed until June 1969 while international loans were agreed. Funds were offered by Germany, France, Italy, Japan, the UK, and the EIB. No country offered enough to build the whole bridge, and because the loans had to be spent by purchasing from the country offering them, it was necessary for international consortia to bid for the job. The winning tender was submitted by Hochtief of West Germany and Cleveland Bridge of the UK. Cleveland Bridge had been part of the joint venture which built both the Forth Road Bridge and the Severn Bridge.

In a 1976 paper to the ICE, Hugh Knox identified the financial arrangements as one of three factors which had a major effect on the way in which the bridge was fabricated and constructed. The other two were the very short time programme and the decision to spin the cables rather than pre-form them.[1]

'Tenderers had to declare with their tender the proportions of the various foreign currencies which they required, and these had to be virtually fixed thereafter,' Knox wrote. Hochtief took responsibility for the substructure while Cleveland Bridge was responsible for the entire steel superstructure, but used subcontractors in other countries to spread the supplies. Most of the UK loan was used to finance the special equipment needed for erection. Most of the West German money was accounted for by the substructure works. Cleveland decided to base its tender on the assumption that most of the steelwork would come from Italy. In the end the towers, part of the suspended structure and the approach viaducts were fabricated there while the rest of the suspended structure, the suspenders and the anchorage steelwork came from the UK, and the main cable from West Germany. Being paid in five different currencies, with different rates of interest and inflation and varying exchange rates, also caused 'complex problems' for the consortium, said Cleveland Bridge's deputy chairman Donald Dixon at an ICE conference in 1978.[2]

The contract start date was 24 April 1970 and the contract period was just 1,020 days – less than three years compared with three-and-a-half to six years for construction of comparable bridges in the past (the Severn Bridge took five-and-a-half years). Because of the steeply sloping shores of the Bosporus with only narrow areas of level ground, and the presence of buildings including an historic palace and its gardens on the Asian side, space and access were limited. This creates particular difficulties with a box section deck because a large amount of space is needed to store completed box sections prior to erection. Little stockpiling of steelwork was possible on site, and most heavy items had to be delivered by barge because of the lack of suitable roads nearby. Locally available electricity supplies were limited, so most major items of plant were diesel powered, with diesel generators used to supply welding equipment, lighting and derricks.

The bridge towers were essentially similar to those of the Severn Bridge. In erecting them, the contractor departed from normal practice by using a crane on each leg of the tower. This proved to have several advantages over the usual single-crane method, notably that the radius at which loads had to be lifted was much smaller, so that the cranes could be lighter.

Only two closures of the Bosporus international shipping lane were needed during the whole contract, to allow the initial hauling lines for cable erection to be strung across.

Learning from the experience of the Severn Bridge, the contractors did not place the box sections directly in the water to float them out prior to lifting into place, but carried them on barges. On the Severn estuary they had proved difficult to handle and immersion had led to problems with the paintwork later.

One incident which could have caused a major setback to the programme occurred in February 1973. A storm in the Marmara sea caused 'unprecedented' waves in the Bosporus. The quay where box sections were being stockpiled, and which had been thought completely sheltered, was overtopped and six box sections weighing up to 40 t were swept into the water. They had to be located by divers and raised by floating cranes. Although they suffered considerable damage, a local shipyard was fortunately able to repair them within three weeks.

The contract was eventually completed on 15 August 1973, 160 days late: this was mostly because of differences in ground conditions at the pier foundations, some design changes and late delivery of some erection equipment which delayed a start on the towers. The bridge cost US$36 million.

Traffic on the bridge surpassed all expectations, with the result that the loans for its construction were paid off from toll revenue within six years. It was built in the urban area of Istanbul, because one of the primary aims was to link the business and residential parts of the city. But the new-found ease of crossing the Bosporus encouraged many more people to live on the Asian side and commute to work, and new housing was built at a great rate. At the same time the bridge attracted long-distance trans-European and Asian traffic, especially freight, causing congestion within the city. According to traffic forecasts, the bridge's three lanes in each direction should have been able to cope with demand to the year 2000. But by 1986 traffic was approaching the maximum capacity of 120,000 vehicles a day.

It was decided to build a second crossing outside the urban area to serve long-distance traffic. Freeman Fox & Partners again won the design commission. The chosen alignment was 5 km north of the first bridge. In association with the Turkish firm Botek, Freeman Fox & Partners also designed 28 km of linking motorway.

As with the first bridge, the towers were placed on dry land. They are located high on the hills adjacent to the Bosporus to avoid disruption of the coastal roads. The bridge is 105 m high with a main span of 1,090 m (16 m longer than that of the first bridge), making Bosporus Two, known as the Fatih Sultan Mehmet bridge, the sixth longest in the world. Other differences include the number of lanes (four instead of three in each direction), and the fact that the hangers are vertical rather than inclined.

The project engineer was Dr Bill Brown, who had worked on the first bridge. But this time he was working not as a partner of Freeman Fox but directly for the Turkish General Directorate of Highways. Freeman Fox continued to provide supervisory staff to work under Brown, who had resigned from Freeman Fox after disagreements over the design of the second bridge. Brown introduced a number of modifications to the original design, causing some concern at Freeman Fox, particularly over the question of who would be liable if anything went wrong.

Brown's philosophy was to make changes where it helped make construction easier: 'When something can be changed to help the bridge's construction, to make the contractor more efficient and quick to everyone's benefit, then it is [changed],' he told *New Civil Engineer* in 1986. He saw his role as being like that of the entrepreneurial engineers of the Victorian age: 'The Engineer should be leader of the project, there to see the engineering is sound and to assist the contractor in every way he can,' he said.[3]

Changes introduced by Brown included modifications to the foundations, a redesign of the cable saddles, and a rethink of deck box welding. One of the most radical modifications was to dispense with side spans altogether and to carry the approach roads on backfill held in place by retaining walls near the tower foundations. The foundations were adapted to suit. This enabled the contractor to bring high-capacity crawler cranes along the approach roads and erect the towers from the ground, dispensing with the usual climbing cranes, the use of which would have made the operation much slower. In theory, said Brown, the towers could each have been erected in eight days, although in practice they took slightly longer.

For the box sections, by reducing the thickness of steel slightly and adjusting the alloy to give a compensating increase in tensile strength, the amount of welding needed on the boxes was greatly reduced.

The contractor for the project was a consortium of Ishikawajima Harima Heavy Industries, Mitsubishi Heavy Industries and Nippon Kokan KK of Japan, Impregilo of Italy and Sezai Türkes Feyzi Akkaya (STFA) of Turkey. A bid led by the UK's Cleveland Bridge & Engineering, Enka Construction of Turkey, Bechtel Overseas of the US and Strabag Bau of West Germany, came second, incurring the anger of the British Prime Minister Margaret Thatcher and the UK Secretary of State for Trade and Industry Norman Tebbit. Mrs Thatcher argued that soft yen loans which supported the winning bid were against an OECD agreement on export credits. The Japanese–Italian–Turkish bid for the bridge itself was actually 10 per cent higher than that of the British bid, but the consortium's overall price of £434 million, including 220 km of approach roads and motorway, was nearly £100 million lower. The consortium's price for the motorway was around £1 million per km less than the British price.

The eight-lane bridge was built even more quickly than the first crossing, opening to traffic six months ahead of schedule on 3 July 1988, after only two-and-a-half years of construction. The associated motorway links included six interchanges and three viaducts, one 690 m long.

Bosporus Two has not reduced congestion on the original bridge by as much as had been expected. The alignment was determined by the need to link up with the Trans-Turkey Highway, part of a trans-European route, but it is too far away from the main city areas to attract much commuter traffic. A new road, the Umraniya to Altunizade highway, is being built on the Asian side to link the two bridges in an effort to correct this. This US$45 million link includes eight prestressed concrete bridges, three precast prestressed viaducts, and two 12-m diameter twin-bore tunnels to avoid disruption of residential areas. Acer is the consultant for the tunnels.

The Galata bridge

The Bosporus bridges may be the most visible, but they are not the only innovative bridges in Istanbul. Completed in 1992 was the new Galata bridge across the natural harbour, the Golden Horn, which replaced the existing early twentieth-century bridge. The original was a pontoon bridge because of the 35-m depth of the harbour. Mott, Hay & Anderson International was appointed to advise on this scheme in 1982 and concluded that the state of the existing bridge, whose steel pontoons had been attacked by the highly polluted waters, was so bad that repairs would be uneconomic.

A high-level bridge was ruled out because it would have ruined Istanbul's historic skyline. Mott, Hay & Anderson designed a new floating bridge, with a lower deck for shops and restaurants like the old one. It had been hoped to make provision in the new bridge for later construction of the metro, but Mott, Hay & Anderson concluded that this would be too expensive. However, by the time tenders went out many potential bidders were keen to provide finance as well. Private finance and alternative designs were allowed as part of the tender provided a conforming bid was submitted as well. Though a floating bridge did prove to be cheapest solutions, a proposal for a conventional bridge from STFA and Thyssen with Leonhardt Andrä & Partners, with West German finance, kept options for the metro open and was not much more expensive. This design was adopted and Mott, Hay & Anderson became proof engineer responsible for checking.

The bridge has six road lanes, plus two for the metro and two footpaths. An 80-m clear navigation opening is provided by two pairs of massive bascule flaps, each leaf being 21 m wide. The decision to go for a fixed bridge meant that it had to be designed against earthquake forces and higher ship-impact loads, since a floating bridge would absorb energy from impact by being displaced. Longer cutwaters supported on piles were provided at the bascule piers, their extra length giving extra stability. Speaking to *New Civil Engineer* in 1992, Joseph

Hart, Mott, Hay & Anderson's project manager, described the decision to support the piers on inherently flexible piles rising through 35 m of water as 'very daring'.[4]

Piling was not without problems. Driving the closed-end tubular steel piles into the normally consolidated clay, the contractor found that refusal occurred much sooner than expected. However, after a wait, pile driving could continue. 'The water in the pores of the clay was being driven out,' explained Hart. This increased its strength. 'What you have to do is let it settle, let the water get back in and recreate those pores before driving can continue.'

The bascule piers are 34 m deep and entirely under water. They contain counterweights which swing into chambers in the piers as the 5,000-t bascules open. Constructing the 15-m deep chambers under water presented difficulties. The solution was to drive a grid of piles up to 90 m long into the harbour bed, standing outside the line of the pier and 8–9 m proud of the water level. Brackets in the walls of the pier projected around the piles so that they could slide down them. The pier could therefore be cast in sections above water-level, then lowered down the piles to allow the next section to be cast on top, and so on.

The bridge deck is comparatively simple, being of in-situ concrete, progressively post-tensioned. The steel bascule sections were fabricated in a small shipyard 30 km away.

The bridge had unexpectedly to be brought into use three months early, in July 1992, after the existing bridge was seriously damaged by fire. While the old bridge was being towed away, one of its pontoons was damaged and partly sank. This raised questions over whether it would be possible to preserve the bridge at a new location as originally planned.

Fortunately all the work necessary for allowing traffic onto the new bridge had been done and it was able to open at once. Control cabins and electronic controls for the bascules' hydraulic equipment were not complete, however. Until they were, the hydraulic valves needed to be controlled 'by the rather more fundamental method of two people simultaneously turning them by hand while in radio contact,' said *New Civil Engineer* at the time.

Public transport

Istanbul is now planning a metro system which will be able to carry 70,000 passengers an hour in each direction. Line 1 will run from 4th

Levent in the north to Yenikapi in the south and then West to Topkapi. Construction contracts have been awarded for the section from 4th Levent to Taksim. For a subsequent phase of the development, a tunnelled crossing of the Bosporus is proposed, to carry both commuter and freight trains. Feasibility studies favour an immersed tube. This is because of the depth of the Bosporus and ground conditions: a rock tunnel would have to be 130 m deep, making the stations at each end excessively deep and creating problems for keeping gradients within the specified limits. A bored tunnel in the overlying sedimentary rocks would still have to be over 50 m deep and would require work to be carried out under excessive air pressures.

South of the Golden Horn a light rail system is also being built. Phase I runs 8 km from Aksaray to Esenier and came into operation in 1990. It was built under a design-and-construct contract by Asea Brown Boveri and Yapl Merkezl. In 1991 the same consortium was awarded the contract for the 10.4-km phase II from Otogar to Gobancesme. The route includes 1.4 km of bored tunnel, 350 m of cut-and-cover tunnel, and 1.3 km of viaduct.

Acer is responsible for reviewing the contractor's designs and supervising construction. It is also advising the municipality of Istanbul on integration of the metro, light railway, tramway and suburban rail networks within the city, and on the options for the rail crossing of the Bosporus.

Sewerage

By the second half of this century Istanbul was in a predicament common throughout the Middle East. Its population had grown at a great rate and no modern sewerage system had ever been provided. Most houses were simply connected to cesspits, holding tanks or open channels, resulting in a serious risk to public health – indeed there were high incidences of hepatitis, polio and typhoid. Such sewers as did exist discharged sewage untreated into the Golden Horn, causing severe pollution.

Studies and proposals for alleviating the situation had been made after World War I and in the late 1950s and early 1960s but no significant progress was made in implementing these schemes. Further studies were carried out in the 1970s and a plan to put them into effect was made in 1984.

There are four areas of water around Istanbul into which sewage could theoretically be discharged: the Black Sea, the Sea of Marmara, the Bosporus (which links the two), and the Golden Horn. The Black Sea is distant from the urban area and the cost of schemes involving direct discharge would be high. The Golden Horn's limited capacity to assimilate sewage had already been exceeded.

There is an unusual pattern of currents in the Bosporus: fresh water from rivers flowing into the strait forms a less dense layer which 'floats' near the surface and flows predominantly south, while a denser layer of salt water below flows in the opposite direction. If an outfall into the Bosporus is deep enough, it will ensure that effluent is discharged into the salt water currents and carried to the Black Sea and dispersed.

Water quality studies showed that wastewater could be discharged at deep level into the Bosporus after preliminary treatment only, but discharges into the Sea of Marmara would have to undergo full treatment first.

Istanbul is divided into 11 major drainage zones, six on the European side and five on the Asian side. The overall plan envisaged seven outfalls into the Bosporus, at Ahirkapi, Baltalimani and Tarabya on the European side and Kadikoy, Uskudar, Kucuksu and Pasabahce on the Asian side. These would all discharge sewage after preliminary treatment only. This comprised phase I of the overall scheme.

Under phase II three outfalls would be built into the Sea of Marmara, at Buyucekmece and Kucukcekmece on the European side and at Tuzla on the Asian side. These would discharge effluent from Istanbul's first biological treatment plants. John Taylor & Sons and Binnie & Partners, in association with the Turkish consultant UBM, were appointed by ISKI, the Istanbul Water & Sewerage Authority, to review and finalise the design for phase I and to provide project management services and supervision of construction. Three firms – Watson Hawksley, with Motor Columbus of Switzerland and Temel Muhendislik of Turkey – were appointed for taking phase II from master planning to tender documentation stage.

In general the approach was to provide separate foul and storm water sewers in areas which originally had none. Limited areas were already served by a combined system and it was decided that there was no advantage in attempting to divide the system. Where there were sewers the basic scheme was to build interceptor sewers along the shores near the existing outfalls to take the sewage to a preliminary treatment works before it was discharged into the Bosporus.

Work began first on the South Halic area, covering 10,000 ha and serving an initial population of two million. Three interceptor sewers totalling 18 km in length were built on the south side of the Golden Horn and on the shore of the Sea of Marmara, discharging to a pre-treatment plant at Yenikapi. They were designed to cope with flows up to the year 2020. About a third of the length was in tunnel, the rest in pipes laid in open-cut trenches. Tunnelling was needed because the interceptors run through densely populated areas with narrow streets, where ground conditions would in any case have made open-cut construction difficult.

Three tunnels, known as Halic, Fatih and Eyup, with 2.2–2.8 m diameters, were driven using three West German Herrenknecht road-heading machines 34 m long. These machines could deal with the wide variety of ground conditions encountered which included sands, gravels, silts, clays and weathered and unweathered rock, sometimes heavily faulted. 'These are the first tunnels in Istanbul. It is not like London where you know you will be working in clay. Here the ground varies enormously and it doesn't matter how many test boreholes you have, you still cannot cover it all. One minute you are in rock and the next in running sand,' Taylor Binnie's tunnel supervising engineer Dick Wearn told *New Civil Engineer* in 1987.[5] Average progress was 40 m a week with a peak of 55 m a week, despite bands of sandstone being encountered which reduced progress to less than 1 m a day. Edmund Nuttall provided technical and operational support to the Turkish contractor STFA on the contract.

The tunnelling machines were equipped with gas detectors which automatically cut the power if they sensed methane. On one machine the detector was triggered bringing the machine to a halt just outside the wall of the ancient city. 'There was a pocket of methane thought to have originated from the garbage of hundreds of years being thrown over the old wall,' explained Taylor Binnie's project manager Jaro Pavel.

The tunnels are up to 40 m deep with a primary lining of 160-mm concrete segments and a secondary in-situ lining 125 mm thick. An unusual feature is a further 1.65-mm thick PVC lining around three-quarters of the diameter. This is to guard against the risk of the sewage becoming septic in summer and giving off hydrogen sulphide gas, which could attack the concrete. Only the top three-quarters of the tunnel is lined because the bottom of the invert will always be submerged.

Operation of the Yenikapi pre-treatment works is fully automatic. The sewage passes through a series of screens and grit removal channels before passing into the effluent pump station and the long sea outfall into the Bosporus at Ahirkapi. Flows up to 6 m^3/s gravitate, and pumping is only needed for flows greater than this. The outfall consists of twin 1,625-mm pipes laid in a pre-dredged trench extending 1.2 km into the channel, where the depth is 60 m.

Later works in the £90 million phase I included the North Halic scheme, with interceptors along the north of the Golden Horn and up the West coast of the Bosporus to a similar pre-treatment works at Baltalimani. On the Asian side interceptors and treatment works were built at Üsküdar and Kadiköy, with outfalls extending 31 m into the Bosporus. Over 100 km of interceptors was built in total. Commissioning of phase I began in 1988.

Phase II, which began in late 1986 and was valued at £200 million, will provide sewerage facilities for the Greater Istanbul area surrounding the central area served by phase I. It includes 150 km of interceptor sewers, two major biological treatment works at Kucukcekmece and Tuzla, and three pre-treatment plants with outfalls into the Bosporus. Population growth up to the year 2020 was taken into account.

Only 18 months was available for design, and a specially modified spreadsheet version of Watson Hawksley's FSEWER sewer programme was developed to be used by the project team based in Istanbul, speeding up the design process and allowing changes during the design period to be accommodated easily.

Three types of area had to be dealt with: well-established developments on the shore of the Sea of Marmara; coastal strip developments consisting mainly of summer residences; and dense mature developments on the shores of the Bosporus. All required different approaches to the design of interceptor sewers.

For the established developments, sewage flows can be predicted readily, trunk sewers already exist and routes for interceptors are well defined. For the coastal strips, most of the holiday homes are built near the beach with no road between them and the sea. In order to get the interceptor as low as possible, so that it could fulfil its function without sewage having to be pumped up into it, it either had to be laid in the beach or the gardens of the homes, or tunnelled. Tunnelling was considered to be the most advantageous option.

For the mature developments, the steep shores of the Bosporus generally meant that the only place to put an interceptor was under the coastal road, which carries heavy traffic. Tunnelling was again the preferred solution – although more expensive it avoided serious traffic disruption.

PVC-lined sewers are again being used, as in phase I, because of the risk of septicity. The biological treatment works which are part of the second phase are the first in Turkey which therefore has little experience of running such facilities. An early decision was to standardise the two works as much as possible to limit the amount of training needed for their operation, and to make staff interchangeable between the two. Developing this idea further, a modular approach was adopted to provide benefits of repetition and familiarity during construction.

A conventional activated-sludge treatment process was chosen because the use of stabilisation ponds, biological filters and land treatment would have needed too much land. Two modules were developed, one comprising a sedimentation unit, an aeration tank and a final settling tank, and the second including units for thickening primary and surplus activated sludge, anaerobic digestion and secondary digestion. One sludge digestion module would be needed for 200,000 people and would serve two sewage treatment modules. Initially six treatment modules and three sludge modules will be provided at each works. A 20:30 standard of effluent is required.

Final effluent will be discharged through a long outfall in order to meet European Community regulations on bathing waters. Sludge, after dewatering, is likely to be disposed of to sea in the short term. In the short to medium term disposal to landfill is being considered, and use in agriculture is the long-term aim. Construction of phase II began late in 1989 and was scheduled to be complete within five years.

Roads

In 1984, Turkey had just 84 km of motorway. By the end of the century it should have over 3,000 route-kilometres, including the world's second-longest suspension bridge. Speaking in January 1992, Onur Kumbaracibasi, the Turkish Minister of Public Works & Housing, said that this highly ambitious programme was part of a strategy to transform Turkey into an 'economic base' to replace its status as a military base during the Cold War years.[6]

Turkey had planned a motorway network in the 1970s but for a decade lacked the political will to implement it. But by 1991, 1,500 km was in use or under construction; a similar length was expected to be completed by 1996 with a further 1,750 km by the year 2000. Ten years later Turkey expects to have a fully integrated network running both north–south and east–west. In pursuit of this goal it had spent

US$5,700 million since the mid-1980s by the beginning of 1993, including US$2,000 million in 1992 alone.

The crucial element in the network is the Trans-Turkey Highway, forming part of the Trans-European network. The Highway will connect Europe with the Middle East. Its runs 3,200 km from Edirne on the Bulgarian border via Istanbul and the second Bosporus bridge to Gerede, north of Ankara, where it splits in two, one leg continuing east to Iran and the other south to Syria and Iraq. Already traffic on this route accounts for 30 per cent of all the vehicle-kilometres travelled in Turkey.

The first phase of Turkey's ambitious motorway programme was let on a design-and-build basis with overseas funding. It included: the 700 km from Edirne to Ankara; the stretch from Tarsus to Pozanti, Gazantiep and the southern ports of Mersin and Iskenderun; and the stretch from Izmir to Aydin and Cesme on the Aegean.

Balfour Beatty was the only British contractor involved, building the Kazanci to Gümüsova section in joint venture with the local contractor Entes. Enka-Bechtel built the Ankara–Gerede section and is also responsible for the Ankara peripheral freeway, the equivalent of London's M25 orbital motorway. Consultants engaged in design or supervision included: Ove Arup, Freeman Fox & Partners, Halcrow, L. G. Mouchel, Rendel Palmer & Tritton, Scott Wilson Kirkpatrick, and Sir Owen Williams & Partners.

A key 'missing link' on the Istanbul–Ankara section is the Bolu tunnel. Construction of the 3-km twin-bore tunnels began in 1991, and involves tunnelling through a complex combination of metamorphosed igneous and sedimentary formations in the north, and a succession of highly altered marls, limestones and breccias in the south. All are intensely faulted because of the proximity of the seismically active North Anatolian Fault. With a cross-section of 12 m by 15 m the tunnels are among the largest in the world to be built in such complex conditions. The New Austrian Tunnelling Method is being used. The consultant is Rendel Palmer & Tritton with O'Sullivan & Graham.

For the second phase of the programme, design and construction are being let separately. This phase will include Istanbul to Izmir, including a 2-km bridge over the Gulf of Izmit. This may be a suspension bridge with a 1,000-m main span or two 500-m cable-stayed spans. Bill Brown is acting as consultant to KGM, the Turkish Ministry of Transport, for the bridge. Also included are Ankara to Pozanti and Gazantiep to Sanliurfa, crucial for the GAP development (see opposite). All three of these schemes were included in the 1993 programme.

Still in the planning stage are the Edirne to Balikesir highway, including a 1,700-m crossing of the Dardanelles at Gallipoli; and links between Ankara and Izmir, and Ankara and Samsun.

KGM general director Atalay Coskunoglu told *New Civil Engineer* in 1991 that continuing overseas involvement was expected. 'Certainly we will need foreign loans; as long as we do, we will employ foreign contractors.' Toll income is expected to pay off the loans 'at the latest within 15 years,' said Coskunoglu.[7]

GAP and the Atatürk dam

The Southern Anatolia Regional Development Project, known by its Turkish acronym GAP, has for some time been seen by some politicians almost as a panacea for the country's economic problems. It is a truly enormous scheme which, if fully implemented as originally planned, could cost US$10,000 million.

GAP is situated in the Tigris and Euphrates river basins and covers 77,000 sq km, a tenth of the country's total area. It includes 22 dams, 19 hydroelectric plants, thousands of kilometres of irrigation canals, plus roads, airports and urban development. Within the overall plan areas are designated to become industrial, agricultural or commercial centres.

The Atatürk dam, on the Euphrates, is the centre-piece of the scheme. The largest dam in Turkey and the fifth largest earthfill dam in the world at the time of its completion in 1991, it will supply water to irrigate an area of 900,000 ha and generate 8.9 TWh/y of energy.

Some have expressed doubts about Turkey's ability to sustain the high level of spending which it has committed to GAP. A master plan study in 1990, when spending was running at US$1 million a day, suggested scaling back and rescheduling the project. A lack of coordination has been noted: upstream of Atatürk, the Karakaya dam and its associated hydroelectric project were almost complete in 1988 – but there was nowhere for its power to go.

Planning for large-scale development of the Euphrates in Turkey can be traced back to the 1930s when hydrological studies were carried out. A plan in 1958 by the Electrical Works Survey Authority recommended a three-dam system for generating energy, in addition to, and downstream of, the separate Keban dam which was completed in 1974. In the 1960s DSI, the State Hydraulic Works, proposed an alternative two-dam system which would also be used for irrigation. In 1970 the most favourable solution was considered to be a three-dam system with irrigation water pumped from the third dam.

The oil crisis of 1973–4 changed the economics in favour of a two-dam system with irrigation by gravity from the second, eliminating the energy requirements for pumping. The two dams were the Karakaya and the High Karababa, renamed the Atatürk to celebrate the centenary of Mustafa Kemal's birth.

Final design of the Atatürk dam was undertaken by Electrowatt Engineering Services of Switzerland in association with Dolsar of Turkey. There were considerable misgivings about the project in Syria and Iraq, which are downstream of the dam and considerably more dependent on the flow in the Euphrates than Turkey is. Turkey has insisted that it will continue to honour its agreement on the use of Euphrates water, under which it must supply an annual average of 500 m³/s downstream, pointing out that it needs to maintain a flow through the dam for power generation purposes.

Syria did not fail to notice, however, that Atatürk was unlike the Keban and Karakaya dams upstream, which were for power generation only. Atatürk is designed for irrigation and, if Turkey wished to do so, it could divert the entire flow of the Euphrates. In fact at the start of impoundment it did completely block off the Euphrates for a month. But Syria and Iraq had been warned, and the flow in the weeks beforehand was increased to boost the average. And tributaries which join the Euphrates downstream were also capable of supplying up to 350 m³/s. Nevertheless, during construction there were allegations of terrorist plots to attack the dam – though experts say that damaging an earthfill dam during construction would be difficult.[8] And, because of concern about cross-border flows, no international finance was made available for construction. Funds for the US$2,500 million project were raised by the Turkish government from the state budget, bonds, and levies on a range of products.

Atatürk is 179 m high, with a crest length of 1.8 km, or 2 km including the spillway. The total volume of the embankment is 84.5 million m³. It is founded on Cretaceous limestones. The foundation surface was minutely inspected for cracks and fissures, and excavated down to absolutely sound rock if necessary. The clay core material was found 4–5 km south of the site; material excavated from the river bed was used in the transition zones; limestone fill was taken from the excavations for the abutments and basalt rockfill was quarried 7 km south-west of the site. The grout curtain is the world's largest, 5 km long and up to 170 m deep.

The Euphrates was diverted during construction, by upstream and downstream cofferdams, into three tunnels, each between 1.3 km and 1.4 km long, which were to become bottom outlets from the dam. The cofferdams were incorporated into the dam's final structure.

The construction contract was won, against international competition and to the surprise of many, by a Turkish consortium of Palet, Seri, and Enerji-su, known jointly as ATA Insaat. One of the largest plant fleets ever seen was assembled for the huge muck shift. Among about 1,000 large machines were 200 Caterpillar 777 dump trucks of 85-t capacity, and another 40 of lesser capacity.

Impounding began in January 1990 after the programme was accelerated by the Turkish government so that power was available sooner than originally planned. The final diversion tunnel was closed seven months early in December 1989, allowing impounding to begin during the winter rather than in the summer of 1990, with the aim of delivering power a year later. To do this, Ferruh Anik, head of DSI, took the calculated gamble that a 200-year flood, which could have inundated the works, would not occur in spring 1990.

Penstocks of 515 m to 640 m in length and up to 7.25 m in diameter supply water to the eight 300-MW Francis turbines of the generating station. The turbines came into operation progressively between 1991 and 1993. From the reservoir, twin tunnels of 26 km feed the irrigation network of the Harran plain.

Turkey today

It might be argued that Turkey's current infrastructure spending is over-ambitious and that targets are unlikely to be achieved. But since 1980, Turkey's exports have increased five-fold, to US$14,900 million in 1992, and despite increases in imports the country has been able to pay its way and finance its debt. One adverse economic effect has been a high rate of inflation, which has affected government popularity.

However, improving the infrastructure to allow the economy to develop is likely to remain a high priority and, despite arrangements designed to ensure transfer of technology to local firms, Turkey can be expected to continue to provide work for the international construction industry.

c h a p t e r   s i x t e e n

THE ROYAL
ENGINEERS

This history would be incomplete without a reference to the British Army's Corps of Royal Engineers (RE). The Corps is the army's centre of military engineering expertise. It provides vital support in war and peacetime, and is responsible for constructing roads, buildings, airstrips, and army camps, and for supplying them with essential services such as water and electricity. These are considerable feats of engineering; they have to be carried out under extreme time pressure, with minimal resources, and in wartime often under fire as well.

The importance of the RE's work is underlined by two examples from World War II. The Corps was involved in repairing severe damage to the quay walls of Benghazi harbour in Libya in 1942, inflicted both by shelling in the fighting and by sabotage by retreating German forces. Until the quays were reinstated the Allied forces were experiencing severe difficulty in unloading vital supplies needed to maintain the Eighth Army's advance on Tripoli. Improvised solutions such as rebuilding the walls with concrete-filled oil drums were used. Sappers always have to be ready to use their initiative and whatever resources are to hand – qualities which stood them in good stead in the Middle East.

The D-Day landings in Normandy were only possible because of the RE's ability to clear the beaches of mines and other obstacles, and the invasion was only sustainable during the subsequent weeks because of their work in developing and constructing the artificial Mulberry harbours and PLUTO – the Pipeline Under The Ocean – enabling men, vehicles, stores and fuel to reach the bridgehead rapidly and efficiently. The RE Brigadier in charge of the Mulberry harbour was Sir Bruce White, who went on to found the firm of Sir Bruce White, Wolfe Barry & Partners which used the techniques developed on the Mulberry harbours in the construction of Dammam port in Saudi Arabia in the 1970s.

Until Britain's decision in the 1960s to withdraw troops east of Suez, Britain had a large military presence in the Middle East. As mentioned above, the strategic importance of the Gulf states had been realised in

Victorian times. The Trucial States, which became the United Arab Emirates, were so called because of the maritime truces between their rulers and Britain, in which Britain undertook to defend them against aggression, while the rulers undertook to stop piracy against British shipping.

Similarly Aden was colonised specifically to establish a military stronghold to safeguard the trade routes to India and the East, and the surrounding provinces became a British protectorate. Britain also had close links with Oman, and provided military support and personnel for the sultan's armed forces after Sultan Qaboos came to power in 1970.

Much of the work which the RE undertook in these times was of purely military significance. But sometimes incidentally and sometimes by design, the RE also did work which benefited the civilian population. Before oil wealth arrived in the Gulf, projects such as basic and simply built roads were exactly what was needed and did much to open up areas of the hinterland.

A good example of this is the Dhala road, built in the Aden Protectorate (now the southern part of Yemen) between 1964 and 1966. The army had been deployed in the area north of Aden, at the request of the Federation of South Arabia, because of aggression by the Quteibi tribe who were backed by Egypt and the Yemen Arab Republic. A particular problem was that the rebels had been laying mines on the main supply route from Aden which ran through Dhala (and was once one of the routes by which spices from India and the East were brought to the land ruled by the Queen of Sheba in ancient times).

Minelaying was a threat both to military convoys and civilian lorries. In 1964, 49 vehicles, including 10 civilian lorries, had been destroyed, and five soldiers and seven civilians had been killed. It was decided to provide bituminous surfacing to the road so that it would be impossible to lay mines without leaving obvious signs.

Between Dhala and Aden, 64 km of road was unsurfaced. The civilian public works department undertook the first 10 km, leaving the rest to the RE. The specification was basic: the road was to be gravel-based with a bitumen and chipping surface dressing; it was to have a 7.5-m wide base and a 4.5-m surfaced width; Irish crossings were to be used at wadis. There were three sections of road: a central section of 11 km following the line of the Wadi Matlah, with more straightforward northern and southern sections. The difficult central section was constructed last. The wadi was steep sided and suffered from severe flooding after heavy rain washed away the existing track which then had to be re-established. For the new road, a line formerly established by occupying Turks along the top of a 10-m high escarpment was followed. The road was completed in 31 months – somewhat longer than originally estimated, largely because it was necessary to mount much heavier guard both on the works and the construction camp.

In the late 1960s the RE carried out several major projects in the Trucial States. This was at a time when only Abu Dhabi had begun to acquire oil wealth. In 1968 one British squadron in the Gulf used plant and explosives to open up an 80 km road suitable for 3-t trucks between Dibba and Khor al-Fakkan along the east coast. According to an RE summary, the squadron 'achieved what was thought to be impossible with existing facilities.' Previously only donkeys and camels had been able to negotiate the ill-defined route because there was no way round the high *jebels* which reached down to the sea.

A detachment of sappers was responsible for drilling and blasting, helped by Arab labourers with picks and shovels. After the engineers had roughly shaped the formation, the Trucial States Development Council (TSDC) undertook grading and surfacing of the road and facing of the embankments. Later, detachments of the RE improved the route so that ultimately the 3-t trucks could travel at 30 mph throughout.

A report in the RE's magazine, *The Sapper*, at the time said: 'The completed road involves some hair raising driving over steep passes, but it opened up that strip of coast, and many people who live along it saw their first vehicle as a result.'[1] Or, as an official report stated: 'The development of the East Coast of the Trucial States is proceeding at a much increased rate as a direct result of the opening of this new route.'

The following year another detachment, using small plant and blasting, improved the route from Dibba inland to Masafi to make it passable for three-tonners. The existing route followed the line of the Wadi Tayibah, where numerous small villages were situated. Many small tasks, such as improvement of irrigation ditches, were carried out as the work passed the villages. In places the wadi was steep sided with precariously balanced boulders, which prevented the use of explosives because of the risk of causing landslides. In these areas, the tough limestone had to be cut away with pneumatic drills. A report records: 'The opening of this track not only improved communication for the local villages, but also simplified patrolling by the Trucial Oman Scouts. Much goodwill was gained with the local population.'

Even more ambitious was the Transpeninsular Highway. This was a black-topped 7.5-m wide road across the peninsula from Sharjah to Fujairah. The initial 80 km from Sharjah to Dhaid was completed by civil contract. Originally the track through the *jebels* was in the bottom of the wadis Siji and Ham, and each year it was washed away by flash floods during the rainy season. The RE worked with the TSDC to select a route above the flood level. Major works included a 10-m deep cutting through a rock saddle, and a stretch of several kilometres where the road had to be cut into the side of the *jebel* rock-face. Four prefabricated bridges, each with a span of more than 30 m, were built, and five multi-barrel concrete culverts were also needed. Work started from a base camp at Bithna in the Wadi Ham in October 1969. Plant, including D8H tractors, graders and compressors, was transported by sea to Fujairah and then taken over land into the wadi. TSDC built abutments for the bridges and was responsible for laying black-top on the road. The road provided for the first time high-speed communication between the Gulf and Indian Ocean coasts of the United Arab Emirates.

Well-drilling was another important activity which did much to benefit the local population. Between May 1968 and September 1970 sappers drilled 11 wells for the TSDC, specifically for the inhabitants of towns and villages in the region. Not all were successful: two were dry and one was saline, but the others produced outputs varying from 90 l/h to 4,700 l/h. In addition two wells were drilled for the ruler of Sharjah, which each yielded 4,000 l/h. With temperatures reaching 54°C, the metal parts of the drilling rigs were too hot to touch with bare hands, and troops were issued with asbestos cooks' gloves to handle them.

Other minor projects included the clearance of an ancient falaj system for the village of Wamm, east of Dibba, and improvement of an airstrip used by both civil and military aircraft at Buraimi Daudi in Abu Dhabi.

In Oman the RE undertook another road scheme, providing a link to seven villages in a fertile bowl in the *jebels* above Rostaq. As with the projects in the United Arab Emirates, communication had formerly been by camel track, in this case along the Wadi Sahtan. Much agricultural produce could be grown in the bowl, and there was a ready market around Rostaq, but it was impractical to transport the produce there. The RE blasted and bulldozed a three-tonner route through the wadi in just over a month in 1969. This was despite being hampered by high temperatures, and the presence of dissident tribes which meant that the camps and working area had to be protected by armed guards day and night.

After the *coup d'état* in Oman in which Sultan Qaboos came to power, the new government wanted to create a favourable impression among the local people in order to remove the main cause of dissent and rebellion, which had been the lack of modernisation and development under the rule of Qaboos's father. Accordingly the new regime was anxious to get short-term development projects under way quickly.

The British army was advising and assisting Sultan Qaboos in modernising his forces, and a number of officers and men from all three British services were seconded to the sultan's armed forces. Early in 1971, an RE detachment was sent to Oman to cooperate with Oman's defence secretary and the British army training team. The sappers helped in several ways, particularly by improving water supplies. Between September and December 1971 eleven wells were drilled in Salalah. At Mirbat, the water supply was improved by constructing catchment channels with a central reservoir and by installing a pump and pipeline to a storage reservoir in the town.

The sultan's forces were commanded at the time by Major-General K. Perkins, who was mainly concerned with ending the rebellion in the southern Dhofar province. He wrote in the Royal United Services Institute Journal: 'The strategy adopted by the Sultanate was based on winning the confidence of the population, the military aim being "to secure Dhofar for civil development."'[2]

Though the inhabitants of Oman were nomadic, the government wished to set up 'centres of contact' with the population as a focus for development. 'To this end, military successes were immediately followed by the construction of access tracks and by provision of well-drilling equipment. The wells became points of focus for the population and were consolidated as government centres by the building of clinics, schools and government shops,' wrote Major-General Perkins.

Sappers were also at work in Sudan between 1972 and 1975. This followed the ending of the civil war by the Addis Ababa agreement between President Nimeiri and the southern forces. With the restoration of peace in southern Sudan, foreign aid and relief organisations began to return. The RE were asked to undertake a project on the main supply route from Wau (at the southern end of the railway from Port Sudan and Khartoum) to Juba (near the border with Kenya and Uganda), 540 km away. This required rebuilding three bridges, two at Tonj and one at Mundri, which was carried out during 1974 and 1975.

The Mundri bridge has a span of 76 m, but the main problems were logistical rather than technical, because of its distance from the

Wau railhead. Work involved removing the old concrete deck and beams, capping the piers with 900 mm of reinforced concrete, replacing five beams on each of the eight spans, and casting a new reinforced concrete deck. Three months were allowed for the job but it was completed three weeks ahead of schedule, which allowed troops to be deployed on other tasks towards the end of the exercise.

'The value of these additional tasks far outweighed the time and effort that went into them, as British soldiers were seen to be helping the community at large,' said an account by Major R. M. Stancombe in a contemporary edition of the *Royal Engineers' Journal*. 'Even the effect of one sapper tradesman undertaking a small job brought rich benefits in terms of Anglo-Sudanese relations.'[3]

These 'subsidiary tasks' included a 1,500-m airstrip which was built for a relief organisation at Amadi; replacement of a 10-m steel and concrete bridge on the Amadi to Mundri road, and two others on the Zaire border. One electrician rewired six wards of the hospital in Wau in a month. The RE also helped with giving Wau, the provincial capital, a face-lift ready for celebration of the signing of the Addis Ababa agreement.

Another benefit was technology transfer. British and Sudanese soldiers worked side by side on the bridge. 'Though there were some language problems,' said Major Stancombe, 'by the end the Sudanese should have been capable of building other such bridges on their own'.

CONCLUSION

In the early 1990s the oil-producing nations were hard hit by the depression in the world oil industry. The oil price dropped to its pre-1973 level in real terms, and sharply reduced the revenues of the oil producers. Even Saudi Arabia and Kuwait, which had seen their reserves drawn down considerably because of the Gulf War, were faced with the unaccustomed prospect of paring away at spending plans.

This is unlikely to be anything other than cyclical, however. Two-thirds of the world's known oil reserves remain in the Middle East, and the importance of the region as the dominant source of energy supply to the West is set to continue well into the next century. Sooner or later the region's oil producers are likely to attain again the dominant position in world oil markets that they held in the 1970s.

The people of the oil-producing states have seen their lives and countries transformed; they have come to expect a high standard of public services and subsidised food, petrol and electricity. But many are nevertheless aware that, compared for example with the newly industrialising countries of Asia and the Pacific Rim, the oil-producing states are lagging behind. Progress in improving education, creating a skilled labour force and diversifying their economies has fallen short of expectations.

There are good reasons for this. Most of the oil-producing states of the Middle East are largely desert, with few natural resources apart from oil, little cultivable land, and small populations. Geographically, they may range in size from city states like Kuwait and the United Arab Emirates to vast countries like Libya and Saudi Arabia, with scattered populations. Their governments have been able to encourage a thriving business sector but growth of industries capable of producing non-oil exports remains limited. This results to a large extent from the lack of resources on which non-oil activities could be based, however willing to invest the governments may have been. But though progress is slow, without the infrastructure provided by the massive construction programmes of the 1970s and 1980s it is doubtful whether it would have been possible at all.

Conversely, Iran, Iraq and Egypt are well endowed with agricultural land and water, and have large populations. But even in these countries resources from which to develop the economic base are scarce. And because of their larger populations, the benefits of oil revenues have had to be spread more thinly. In these circumstances governments have tended to focus the benefits onto their own power-base. This was a factor in the Iranian revolution.

Indeed many have seen Islamic fundamentalism as a potential solution to Arab problems. Supporters of this tendency reject attempts to emulate the West and argue that a strong Islamic state would be better able to resist western domination.

An alternative view of the future is that the Middle Eastern states need to become more liberal and democratic. As the size of the educated middle class grows it is likely to be increasingly frustrated by its lack of influence on government. In fact, many of this growing educated élite have chosen to live and work abroad.

There has been some progress towards greater democracy in Egypt and Jordan. Saudi Arabia has set up a *shura*, its first consultative council for 60 years. Somewhat faltering steps have also been taken in Kuwait. The elected assembly, reinstated after the Gulf War, has enthusiastically addressed its task of scrutinising government actions, compounding the problems the government already faced because of reduced oil revenue. On past experience such experiments have ended with elected assemblies being dissolved when they became too troublesome, and Kuwait's reaction will be instructive.

A repetition of the factors which provided so much need for the services of construction firms is unlikely to be seen in the Middle East or anywhere else. There is still work to be had in the oil states, notably the United Arab Emirates, Saudi Arabia, and Qatar (as it develops its natural gas resources). Long-term developments like the Greater Cairo Wastewater Project will continue. But the basic infrastructure of the region is now in place, and the emphasis has changed to renewal and refurbishment of oil production facilities which are nearing the end of their life.

Meanwhile, competition is much more intense than in the 1970s. Depressed oil prices will mean that work goes ahead more slowly: in Iran, the expected upturn in work has almost ground to a premature halt because of lower than expected revenues. Elsewhere in the world, the developing countries of Asia can be expected to be more insistent about technology transfer when they bring in overseas help, in order to gain the required expertise quickly for themselves.

But the achievements of the engineers who helped to transform the Middle East should not be underestimated. In just a few years a succession of mainly poor and undeveloped countries were provided with schools, hospitals, electricity, safe drinking-water and sewerage systems, plus the infrastructure to underpin economic development. As in Victorian Britain, the impact on the health, education and overall quality of life for the general population has been overwhelmingly positive.

NOTES

Chapter 1

1. Sir Harold Hartley, 'Engineering and civilisation', from 'Engineering societies in the life of a country', Lectures Commemorating the 150th Anniversary of the ICE, ICE, London 1968.
2. Sir Harold Hartley, 'The engineer's contribution to the conservation of natural resources', The Graham Clark Lecture 1955, *ICE Proceedings* Pt 1, Vol. 4, September 1955.
3. G. Roux, *Ancient Iraq*, London, George Allen & Unwin, 1964.
4. D. G. M. Roberts, 'Wealth, health & sport', Inaugural Lecture as Royal Academy of Engineering Visiting Professor in the Principles of Engineering Design, Loughborough University of Technology, 5 May 1993.

Chapter 2

1. Farahnaz Pahlavi Dam: Tehran Regional Water Board brochure.
2. R. Marwick and J. P. Germond, 'The River Lar multipurpose project', *Water Power & Dam Construction,* April/May 1975.
3. *NCE* 26, April 1990.
4. *NCE* 29, November 1979.

Chapter 3

1. A. M. Hamilton, *Road through Kurdistan*, London, Faber & Faber, 1958.
2. 'Middle East change', *NCE* Supplement, July 1979.
3. 'The impact of the war on Iraq', Report by a mission led by Mr M. Ahtisaari, UN Under-Secretary General for Administration and Management, Pamphlet No. 7, *Open Magazine Pamphlet* Series, Westfield, New Jersey, April 1991.
4. *NCE*, 19 February 1976.

Chapter 4

1. 'Arab World', *NCE* Supplement, 30 June 1977.
2. *NCE*, 19 June 1975.
3. G. R. D. Marshall and R. Channa, 'The Manama-Sitra Causeway Bridges – replace or repair: The engineering and economic assessments', Second International Conference on the deterioration and repair of reinforced concrete in the Arabian Gulf, Bahrain, October 1987.

Chapter 5

1. H. W. Try and M. A. F. Rush, 'The experiences of a medium sized construction company in the Middle East', *Management of International Construction Projects*, London, Thomas Telford Ltd, 1984.
2. NCE Saudi Arabia special feature, 17 June 1982.
3. 'Arab World', *NCE* Supplement, 30 June 1977.
4. *NCE* Supplement, 8 July 1976.
5. L. Grunt, 'Trial structure at Tours', *Arup Journal*, Sept 1971.
6. *Op. cit.*
7. *NCE*, 21 June 1979.

Chapter 6

1. 'Arab World', *NCE* Supplement, 30 June 1977.
2. Threlfall, Ryder & Wood, 'The Sahlia complex, Kuwait', *Arup Journal*, July 1979.
3. J. P. Cowan and P. R. Johnson, 'Re-use of effluent for agriculture in the Middle East', *Reuse of Sewage Effluent*, London, Thomas Telford Ltd, 1985.

Chapter 7

1. 'Arab World', *NCE* Supplement, 30 June 1977.
2. C. Hogg & G. M. Swan, 'Water supply for Doha, Qatar', *ICE Proceedings* Vol. 7, Aug 1957.
3. *NCE*, 15 January 1976.

Chapter 8

1. Innes & Farthing, 'Al Ain International Airport – planning and contract strategies', *Airports and Automation*, London, Thomas Telford Ltd, 1992.
2. M. C. Bailey, 'Sewerage and sewage treatment in Dubai – an overall view', *Middle East Water and Sewage*, July/August 1980.
3. Leigh *et al.*, *125 Years of Halcrow*, Reading, Sir William Halcrow & Partners, 1993.
4. Daniels & Sharp, 'Dubai Dry dock: planning, direction and design considerations'; Cochrane, Chetwin & Hogbin, 'Design and construction', *ICE Proceedings* Pt 1, Vol. 66, Feb 1979.
5. P. D. V. Marsh, 'The Dubai aluminium smelter project', *Management of International Construction Projects*, London, Thomas Telford Ltd, 1984.

Chapter 9

1. Sultan Qaboos University, *Construction News Achievement Award* (supplement), 1987.

2. Leigh *et al.*, *125 Years of Halcrow*, Reading, Sir William Halcrow & Partners, 1993.

Chapter 10

1. J. E. G. Palmer & H. Scrutton, 'The design and construction of Aden oil harbour', *ICE Proceedings,* Pt 1, Vol. 5, July 1956.
2. M. R. Lane, *The Rendel Connection*, London, Quiller Press 1989.
3. L. Jackson, 'Aden: The planning and construction of a new township on a desert peninsula', Conference on Civil Engineering Problems in the Colonies, ICE, June 1956.

Chapter 11

1. 'Middle East nations on the verge', *NCE* Supplement, 17 April 1978.
2. 'Roseires Dam; Ministry of Irrigation and Hydro-Electric Power, Republic of Sudan', brochure commemorating opening, 1966.
3. *Op. cit.*

Chapter 12

1. 'The Great Man Made River Project', The Management and Implementation Authority of the Great Man Made River Project, 1989.
2. M. Simpson, 'Benghazi City Roads Project', *Arup Journal*, July 1980.
3. C. Wade, 'Al- Fatah University Libya', *Arup Journal*, September 1978.

Chapter 13

1. W. Willcocks, *Sixty Years in the East*, Edinburgh and London, Blackwood & Sons, 1935.
2. Obituary, *ICE Proceedings*, Vol. 8 September 1957.
3. 'High Aswan Dam: Vital achievement, fully controlled', International Commission on Large Dams 61st executive meeting and symposium, Egyptian National Committee on Large Dams, Cairo, November 1993.
4. J. A. Allan, 'The impact on Egypt of the High Dam at Aswan', School of Oriental and African Studies, University of London, Paper presented to the British-Egyptian Society, 25 November 1993.
5. D. G. M. Roberts and E. Flaxman, 'Greater Cairo Wastewater Project: history, development and management'. *ICE Proceedings,* Pt 1, Vol. 78, August 1985.
6. *NCE,* 28 November 1991
7. *Ibid.*
8. *NCE,* 12 December 1974.

Chapter 14

1. J. E. G. Palmer, 'Development of the Port of Aqaba', Conference on Civil Engineering Problems Overseas, ICE London, 1960.
2. 'Middle East nations on the verge', *NCE* Supplement, 27 April 1978.

Chapter 15

1. H. Knox, 'Bosporus Bridge – construction of superstructure', *ICE Proceedings*, Pt 1, Vol. 58, November 1975.
2. D. Dixon, 'Managing an overseas contract in a developed country – Bosporus Bridge', Management of Large Capital Projects, ICE, London, 1978.
3. *NCE*, 11 December 1986.
4. *NCE*, 13 August 1992.
5. *NCE*, 26 March 1987.
6. World Highways September/October 1993.
7. *NCE*, 18 July 1991.
8. *NCE*, 27 November 1986.

Chapter 16

1. 'Engineering in the Trucial States: 10 Field Squadron in the Persian Gulf', *The Sapper*, May 1969.
2. Maj. Gen. K. Perkins, 'Oman 1975: The year of decision', *RUSI Journal*, Vol. 124, 1979.
3. Maj. R. M. Stancombe, 'Sappers in Sudan 1972–75', *RE Journal*, 1976.

INDEX